ASH

Reaktion's Botanical series is the first of its kind, integrating horticultural and botanical writing with a broader account of the cultural and social impact of trees, plants and flowers.

ASH

Edward Parker

REAKTION BOOKS

For Susannah

Published by
REAKTION BOOKS LTD
Unit 32, Waterside
44–48 Wharf Road
London N1 7UX, UK

www.reaktionbooks.co.uk

First published 2021
Copyright © Edward Parker 2021

Printed and bound in India by Replika Press Pvt. Ltd

A catalogue record for this book is available from the British Library

ISBN 978 1 78914 356 0

Contents

A large ancient European ash tree in a meadow in Devon.

Introduction

꧁

Ash are a cosmopolitan group of trees belonging to the genus *Fraxinus*, and today, in their various forms, they are widely distributed across the temperate areas of the northern hemisphere, as well as in a few parts of subtropical Asia and North America.

In many ways ash trees are physiologically quite unremarkable, comprising a tall, slender, straight trunk that supports a graceful dome of feathery leaves. They are not particularly tall or stout. They do not sport flamboyantly coloured flowers or, despite living to great ages, display the ostentatiously ancient-looking, gnarled trunks of trees such as the olive (*Olea sp.*) and yew (*Taxus sp.*) are known for. They tend to just quietly go about their business, blending into the woodland communities and the agricultural landscapes they inhabit, and go largely unnoticed in towns and cities. Even great writers and artists have failed to immortalize the ash tree. Writers such as William Wordsworth and John Clare, despite their enduring love of trees, hardly mention the ash at all, and among the 884,647 words of the complete works of Shakespeare the ash tree makes a guest appearance only once:

> . . . let me twine
> Mine arms about that body, where 'gainst
> My grained ash one hundred times has broke
> And scarred the moon with splinters.
> (*Coriolanus*, IX.5)

7

Similarly, in his book *Sylva Britannica* published in 1722, the tree artist Jacob George Strutt, out of the many arboreal wonders he chose to feature, displayed only two grand old ash trees, the largest of which was the 'Great Ash of Cannock' in Stirlingshire, Scotland. However, despite their rather unremarkable, almost generic appearance, few trees have had such an impact on the development of human civilization over the last 5,000 years as the ash tree.

The ash tree is known to have been central to the lives of many ancient Indo-European cultures such as the Greeks and Celts, as well as Norse and Germanic tribes, for thousands of years. So much so that over time the ash tree took on a mythical status. In Norse culture, for example, the first man was believed to have been fashioned from an ash log (woman from elm),[1] while in Greek mythology the third race of humans were said to have been created from ash trees by Zeus.[2] Even Zeus himself was thought to have been tended by 'ash nymphs' known as *meliai*,[3] and fed manna, the waxy sap that exudes from the trunks and branches of some species of ash. Ash sap was still being used as an infant's first food in locations such as Scotland well

Various species of *Fraxinus* contribute to the biodiversity of the great forests of North America.

The Great Ash of Cannock in Scotland, as represented by Jacob George Strutt in his *Portraits of forest trees, distinguished for their antiquity, magnitude or beauty* (1830).

into the nineteenth century,[4] and has been commercially harvested on Sicily for use in medicines for more than 2,000 years.

For the Norse, Celtic and Germanic cultures the ash was so important that it was believed that a single infinitely large ash tree provided the cosmic infrastructure on which the whole of life and the universe was supported, both literally and metaphorically. In all three cultures it features as the Tree of Life, or World Tree. In addition to its role as the World Tree and as the progenitor of mankind, in Celtic culture the ash tree was often also associated with magic and superstition – much of the paraphernalia of witches, wizards and Druids, such as wands and staffs, was said to have been crafted from ash timber. Ash was also one of a trinity of trees required to form sacred groves in Irish culture. Across the Atlantic Ocean ash trees were culturally important to a host of native people in North America.[5] For example, the green ash (*F. pennsylvanica*) was, and remains, sacred to the Omaha and Ponca peoples.[6]

Orange and black figures depicted as on pottery amphora, showing Achilles fighting in the Trojan War.

The question is: 'Why was the ash so significant in the mythology and culture of so many peoples?' The answer probably lies in its incredible utility. It is difficult to ascertain when the ash tree first became useful to people, but it is possible that the use of ash wood extends back thousands – if not tens of thousands – of years, to a time when humans were first beginning to control fire. Ash timber has a number of particular qualities that may well have helped facilitate the spread of this new game-changing technology. The most important of these in relation to fire is that its timber has an extremely low moisture content when compared to other similar types of trees. This means that the wood will season (dry) much more quickly than that from other trees and will even burn when still green; a potentially life-saving quality not lost on ancient peoples.

There is evidence that wooden artefacts were already being manufactured as long as 400,000 years ago.[7] Unfortunately, as ash timber rapidly decomposes in damp conditions, much of the possible evidence of the ancient use of ash wood in artefacts that may have been found on archaeological sites has long since disappeared. However, in exceptional circumstances, such as in the extraordinarily dry conditions of the burial chambers of Egyptian pyramids, ash artefacts, such as components of chariot wheels, have been found intact, hinting at a long history of not only usage but an ancient international trade in the timber.[8]

Through the study of both archaeological and linguistic evidence it is known that the ash spear largely replaced hunting spears made of yew (*Taxus baccata*) of the Palaeolithic era during the Bronze Age. Ancient peoples were thought to choose ash as the material with which to make their spears because of its tough, impact-absorbing and straight-grained timber. Formidable ash spears were said to have been wielded by Achilles and Odin in Greek and Norse mythology respectively. There is also evidence that ash played a vital part in facilitating settled agriculture, with ash wood becoming one of the main materials used in the manufacture of many vital agricultural implements such as ploughs and rakes. It was particularly suited to the manufacture of tool and weapon handles. Even its harvested foliage provided an important fodder crop for domesticated livestock,

Silhouette of branches and leaves of *F. angustifolia* on the island of Sicily.

particularly in higher elevations of locations such as the Alps and Atlas Mountains.[9]

By the Middle Ages in European countries such as Britain, ash trees were crucial to the economy because of the sheer number of purposes to which the timber and other parts of the tree could be put.

From the Middle Ages well into the twentieth century, European ash trees provided the raw materials for many crafts and industries of major economic importance. For example, the frames and composite wheels of the fastest and most comfortable coaches in the nineteenth century were made from ash timber, with an estimated 1.5 million coaches employing ash made in the USA in 1900.[10] In addition to the sheer utility of the ash timber, its ability to be cut back regularly under management systems known as 'coppicing' and 'pollarding' meant that large quantities of small-gauge portable poles were available, to be used for everything from firewood to arrow shafts and tool handles to furniture.

The pale timber of a freshly cut ash log.

Distinctive upturned twigs and spear-shaped leaves of ash in Britain during springtime.

The leaves, bark and samara (winged seeds) also provided a veritable pharmacopoeia of traditional remedies right across the world for thousands of years. In Europe, for example, the leaves were often used as an alternative to 'Peruvian bark' (quinine) in the treatment of fevers and agues such as malaria.[11] In North America various native peoples have been recorded as using extracts from native ash trees for the treatment of medical complaints, such as earache and snake bites, while the medicinal use of extracts of *Fraxinus sp.* in Greek and Chinese culture has been recorded for more than 2,000 years. Today scientific researchers are investigating the range of complex chemicals present in the various species of ash trees and their potential efficacy in the treatment of diseases such as HIV and Parkinson's.[12]

Sadly, most people today would probably find it difficult to describe the general identifying features of an ash tree. Considering that ash trees in the northern hemisphere are numbered in their billions it is surprising that so few people are familiar with what one might look like, and even more so considering that the ash is still widely used in house construction, furniture-making and the manufacture of sports equipment such as baseball bats. It is also less than a hundred years ago that ash was still being used in the construction of what could be considered quite modern products such as cars, trains, buses and even aeroplanes.

In understanding just how central ash once was to the life of people, particularly in Europe, it is almost shocking to witness how the ash tree largely fell from the collective memory of modern society during the second half of the twentieth century. Today, for many, it simply represents a generic woodland tree with no more or less value than other similar-looking trees. Even worse is that, because of its prodigious ability to regenerate and colonize, it is even considered a weed in some quarters.

After a relatively short period of general lack of interest, ash trees are beginning to enter the minds of people once again. This time, though, rather than acknowledging or celebrating their intrinsic usefulness or their ecological importance, it is the concern about their imminent catastrophic decline: a predicted biological Armageddon where billons of ash trees across the northern hemisphere are likely to perish at the hands of several diseases unwittingly spread by humans. In a devastating pincer movement ash trees are falling victim to pests and diseases, such as ash dieback in Eurasia and the actions of the pupae of the emerald tree borer in North America. Based on what has already been witnessed in Eastern Europe and North America, an environmental catastrophe of greater severity than that caused by Dutch elm disease is unfolding.[13] The environmental impact could be enormous. The very nature of some of the great northern forests will change forever, and some wildlife particularly reliant on ash for habitats could virtually disappear over the next couple of decades in certain areas of the world. It is a cruel twist of fate that the very trees that helped facilitate much of our technological development, and which were so vital to the livelihoods of millions of people for so long, could be largely wiped out, in part by our own short-sighted actions.

The majority of ash species (such as *F. pennsylvanica*) take the form of a medium-sized tree with a slightly asymmetric canopy

one

The Botany of *Fraxinus*
❧

A sh (*Fraxinus*) is just one of 24 distinct genera belonging to the global family Oleaceae, which includes olives, lilacs and privets, and which has its origins some 200 million years ago. Within the Oleaceae family the genus *Fraxinus* is unique in two major ways – having relatively large compound leaves with oddly opposed leaflets and a single-seeded samara (winged seed).

The genus was first described by Carl Linnaeus (1707–1778) in 1753.[1] However, because of many confusing similarities in morphology between, and within, species, more than eight hundred taxa (species, subspecies, races and so on) have been described over the last 250 years. Today 48 true species have been described in the academic literature.[2] Much specific research work has been undertaken on *Fraxinus*, because of the economic importance of several species such as *F. americana* and *F. excelsior*. However, the latest and most thorough monograph of the entire genus, a publication by A. Lingelsheim, dates from one hundred years ago.[3]

Ashes (*Fraxinus sp.*) occur across the northern hemisphere. They are particularly abundant in the temperate zones between sea level and 1,000 m (0.6 mi.) but do not extend into the Arctic.[4] A few species also occur in the arid zones and subtropics of North America and Asia.

Fraxinus Morphology

Species in the genus *Fraxinus* mainly comprise tall or medium-sized trees with straight trunks, some of which are able to grow to heights in excess of 40 m (130 ft), as in the case of two of the most commercially valuable species, *F. excelsior* and *F. americana*. However, as with many aspects of *Fraxinus*, there are several exceptions to the rule. *F. gooddingii*, for example, is small and multi-stemmed and is more shrub-like in appearance. Ash trees have a relatively open, asymmetrical canopy of feathery leaves. This openness of the canopy, and the fact that ash species generally come into leaf later than many of the trees among which they grow, permits the development of a diverse understory.

In many ways, the diverse morphology of ash species does not allow a one-size-fits-all description to be employed. While there are many similarities between species that do allow them to be compared

Ash trees have a more open canopy and come into leaf later than many northern hemisphere trees, allowing a rich understorey of woodland plants to thrive in the spring.

there are also a number of exceptions to the rule. Generally, the main structures that enable *Fraxinus* species to be identified in the field and in the fossil record are their leaves, seeds and flowers.

Leaves

Ash species are generally recognizable by their leaf morphology. Most species of ash have compound leaves that are described as imparipinnate. This means that their compound leaves comprise a number of pinnate (spear-shaped) leaflets set on a central rachis (stem), which are oddly opposed and have one final leaflet at the tip. There are generally between three and eleven tooth-edged leaflets on each rachis (central stem). These can vary greatly in size from 1 cm (0.4 in.) in some species, such as *F. gooddingii*, to between 7 and 13 cm (3–5 in.) in species such as *F. quadrangulata*. As with many aspects of ash there are a number of exceptions to this rule, such as *F. latifolia*, which has leaflets that are more oval than spear-like in shape, and *F. anomala*, which has simple non-compound leaves. Most ash species are deciduous, with the occasional exception such as the small semi-evergreen tree *F. griffithii*, for example, which can be found in countries such as Japan and the Philippines.

There is much variation regarding leaf morphology both between species and within species, including the length and shape of structures such as the leaflets and how indented or toothed the margins are. There are considerable observable variations at a microscopic level too, where structures such as stomata and epidermal papillae (hairs on the leaf surface) can vary considerably. It is these small differences between species and within species that have given rise to most synonyms (trees which appear to be of different species/races/varieties but are in fact simply permutations within a single species) and the extraordinarily high number of *Fraxinus* taxa described over the last couple of centuries.[5]

Flowers

Another highly identifiable feature of species of the genus *Fraxinus*, and one of the features that enable species to be distinguished from each other, is their flowers. The flowers of ash species tend to be borne in clusters on panicles (occasionally racemes), which generally emerge with the leaves from terminal or lateral buds. These can be quite showy, as with the blousy off-white flower clusters of the popular ornamental tree the manna ash (*F. ornus*). On closer inspection, the tiny individual flowers that make up these clusters are among the simplest forms of inflorescence of any angiosperm, often comprising just two stamens and one pistil.[6]

Ash trees such as *F. pennsylvanica* have distinctive spear-shaped compound leaves.

These remarkably simple flowers can take on one of three subtle variations in form, termed 'complete', 'incomplete' and 'naked', the morphology of which is largely linked to the mechanism of pollination. 'Complete' flowers are distinguished by having two to four small individual white petals (forming the corona) and a cup-shaped, dentate calyx surrounding either stamens or ovaries. Ash species that display this 'complete' form of flower are entomophilous, that is, that they are insect-pollinated, such as *F. chinensis*. Species that have 'incomplete' flowers retain the dentate calyx but no longer have the simple petals (corona). Many of the species with 'incomplete' flowers are entomophilous, but some have partially adapted to be anemophilous (wind-pollinated) too, as with *F. gooddingii*. 'Naked' or apetalous flowers have neither the petals nor the calyx and are almost entirely wind-pollinated, such as *F. velutina*.

There are a variety of pollination and breeding systems operating within the genus *Fraxinus*. Approximately one-third of *Fraxinus* species are insect-pollinated and two-thirds wind-pollinated.[7] The sexual reproduction of ash trees also varies between species in the genus. Most ash species, for example, are dioecious (with male or female flowers on separate trees) or polygamous (some flowers with stamens only, some with pistils only and some with both). However, some of the most recently evolved species, such as *F. excelsior*, can have male, female and hermaphrodite flowers on a single tree, while others are what is termed androdioecious (having male and hermaphrodite flowers only).[8] This is a relatively rare breeding system, not often found outside the Oleaceae family.

Samaras (Winged Seeds)

Another distinctive feature of ash species is the clusters of winged seeds (samara) that are more popularly known as 'keys'. Each comprises a single flattened seed attached to a papery wing. In the case of *Fraxinus* the samara have a bilateral symmetrical wing (rather than one with a thickened leading edge such as in *Acer* species, that is,

All ash species have distinctive winged seeds known as samara.

maple and sycamore trees). When they fall from the tree they travel a short distance with the heavy end down until the wing starts to rotate. This auto-rotation slows the rate of descent, allowing the seed to be carried horizontally away from the tree by the force of the wind. These winged seeds can be dispersed over impressive distances as they helicopter on the wind, and the samara from some ash trees have been recorded as travelling more than 200 m (655 ft) from their parent tree. The system of seed dispersal is very effective in ash trees.[9]

In experiments it was noticed that despite the fact that *Fraxinus* samara have a much higher terminal velocity (they fall more quickly to the ground) than that of samara of *Acer* species, they were still recorded to have travelled over longer distances than *Acer* seeds.[10] In studies the ash samara were recorded as travelling more than three times the distance away from the parent tree as that of similar winged seeds produced by *Acer* trees. The determining factors for the distance of dispersal are a combination of the height from which the seeds are released and the strength of the wind. The reason why *Fraxinus* samara travelled further from the parent tree was not because of size, weight or wing length but more a function of over how long a period the samara were released. Tree species such as *Acer* tend to release their samara in the autumn only, whereas *Fraxinus* release their samara during the autumn, winter and spring, allowing them to take advantage of winter storms to help disperse the seeds further.[11]

Generally, once the seeds have fallen to the ground they lie dormant for a period before germinating. Young seedlings in a woodland situation can, if necessary, lie dormant for a number of years before exploiting a gap in the canopy. Areas of the forest floor in European, North American and Asian broadleaved woodland where the canopy has been breached, by either natural events or human activity, can quickly become carpeted with young ash seedlings that vigorously compete to plug the gap and reach the light. Ash seedlings of trees such as *F. excelsior* are particularly well adapted to take advantage of field margins and disturbed land.

Trunk, Branches and Bark

The trunks of most ash trees are straight-growing, although some of the species such as the *F. gooddingii* are more shrub-like with multiple stems. The bark of most species when young is smooth. However, as ash trees age the texture of the bark becomes coarser, eventually forming a characteristic latticework of diamond-shaped furrows. The bark of trees in the genus *Fraxinus* generally contains a high level of cellulose as well as a number of complex chemical compounds including coumarins, secoiridoids, phenylethanoid glucosides, phenolic compounds and flavonoids, a number of which are used as a chemical

Small shrub-like *F. gooddingii* ash trees are found as far south as Mexico and Central America in arid conditions, such as on Isla Tiburon in the Sea of Cortez.

defence against predators and grazers.[12] Many of these chemicals have also been shown to have medicinal and other properties useful to humans. The bark of trees such as the blue ash (*F. quadrangulata*) also yields dyes, which have been used by America's indigenous peoples to colour both basketry and fabric for millennia.[13]

The trunks of some ash trees, such as *F. americana*, *F. pennsylvanica* and *F. excelsior*, can grow to substantial size. The trunk of the largest current common or European ash (*F. excelsior*) is said to be located at Lourinha, São Bartolomeu dos Galegos, in Portugal, with a girth of around 13 m (42 ft), and is believed to be between eight hundred and nine hundred years old.[14] In exceptional circumstances *F. excelsior* can grow to a height of 45 m (148 ft). In Ireland there is another giant European ash tree with a girth of 10.77 m (35 ft) located in a private park near Thurles, but this is likely to have been a multi-stemmed tree formerly.[15] The largest recorded white ash tree (*F. americana*) is located in the botanical garden of the University of Vienna, Fasanviertel, in Austria, and measures 6.18 m in girth (20 ft).[16] The largest white ash

The bark of young ash trees is smooth but as they age it becomes more fissured.

recorded in North America is located on the George Washington victory trail, near West Trenton, and has a girth of 6.18 m, and an estimated age of 350 years.[17]

The timber of ash trees is known as ring-porous. This is where the laying down of the new timber growth each year, which appears as annual rings, is characterized by visible differences in the growth and structure of the new sapwood between the early spring and the late summer part of the growing season. The early season growth is faster and characterized by the development of large vessels (xylem and phloem) within the new timber, often large enough to be visible to the naked eye. In contrast, the timber is laid down more slowly during the latter part of the growing season, making it denser and characterized by smaller vessels.

The branching of the shoot in young European ashes is known as monopodial; this is where the tree prioritizes vertical growth over lateral branching. It tends to be self-pruning (dropping lower branches) when young, helping divert energy into vertical growth. The side branches are generally opposed. The group of North American trees that are believed to be the earliest forms of *Fraxinus* display a curious rectangular cross-section to their branches and twigs. For example, *F. quadrangulata* is typified by four corky ridges on its branches.[18]

The terminal winter buds that form at the tips of the canopy branches in most *Fraxinus* species are characteristically matte black and packed with brown hairs, and tend to have three pairs of protective modified bud scales.

Roots

Ash trees generally develop a far-reaching, rope-like root system that extends through the uppermost horizons of the soil and is particularly prolific in the top 5 cm (2 in.) of the soil profile.[19] Even smaller trees have extensive root plates that can rapidly impoverish the surrounding soil, a feature that has been noted by farmers for millennia

– so much so that the Anglo-Saxons believed that crops would wither and perish in the shade of an ash tree.[20] However, the development of the root plate can vary depending on the conditions under which the particular trees are growing and, most importantly, the presence of light and shade. The tenacious roots of ash trees enable them to colonize slopes with gradients of up to 80 per cent (72 degrees).[21]

Regeneration

Unlike some trees, such as aspen and lime, ash trees are not clonal. This means that they are not able to produce new genetically identical trees via suckers off the main trunk. However, ash species generally respond well to both natural and human-induced coppicing and pollarding. Coppicing is where a tree is cut close to the ground in order to produce a harvest of poles and foliage every three to twenty years. Pollarding is similar but varies in that the tree is cut just above the browse line of wild and domestic animals, 2–3.5 m (6½–11½ ft) above the ground. Coppicing and pollarding both take advantage of the ash tree's natural response to its trunk or branches being snapped or cut, where new buds are stimulated to bore their way through the bark and form the next generation of shoots. Coppicing and pollarding are human activities that have been shown to date back to Neolithic times.[22]

Evolution

Today the total number of species in the genus *Fraxinus* is generally agreed to be 48, which are in turn grouped into six distinct sections.[23] It is also generally agreed that the initial dispersal event of *Fraxinus* from North America into Asia is believed to have occurred during the Oligocene (33.9–23 MYA), followed by another major intercontinental dispersal event where Asian species expanded west, creating the new Eurasian section of *Fraxinus* during the Miocene (23–5.3 MYA) and then finally a single species migrating back into North America.[24]

The oldest ash species fossil so far discovered was found in North America, in rocks that were laid down during the Eocene epoch, which occurred between 55 and 34 million years ago.[25] It was unlike anything found in earlier deposits, displaying the distinctive flowers, samara and compound pinnate leaves by which we identify *Fraxinus* today. The flowers of the specimen found in the oldest rocks were 'complete', having a calyx and corolla, indicating that the very first *Fraxinus* was likely to have been pollinated by insects.[26]

One of the most obvious evolutionary changes associated with ash trees is the change over time of the morphology of their flowers. Over millions of years, what were already relatively simple flower structures have become even simpler, with the loss of the corolla in some species and the loss of both corolla and calyx in others. This would indicate that the species with the simplest flower structures are likely to have evolved to become wind-pollinated. However, the evolution of flowering plants and trees such as ash is not a conveniently predictable process and some species over time have either partly or fully reverted to complex flowers again.[27]

Not all distinctions between species are obvious, and therefore researchers, when looking into the evolution of particular plants, cross-reference the morphological evidence with comparisons at a molecular level using techniques such as molecular dating. Research suggests that the first of the new genus *Fraxinus* appeared in south-eastern North America around 40 million years ago.[28] This unique species then relatively quickly diversified into two more species, forming the initial section *Dipetalae* – a 'section' being a subdivision of a genus into which species can be grouped. More species belonging to two further sections evolved also on the North American continent: the *Melioides* and later *Pauciflorae*. During the Oligocene (33.9–23 MYA),[29] at times when the sea level had fallen sufficiently for land bridges connecting North America and Asia to appear, ash tree species were able to migrate into Asia. They eventually became isolated as the sea level rose again, leading to the creation of two more sections of *Fraxinus*, namely *Ornus* and *Sciadanthus*. In periods of

An ash leaf fossil found in south-central North America.

stable warm conditions the species in the Asian section, *Sciadanthus*, were able to expand once more, but this time right across Asia and into parts of Europe and North Africa, leading to the latest section – confusingly also called *Fraxinus*.

Whereas all the earlier sections were very much geo-specific in that they grouped together in geographical clusters, the most recent species that form the latest section (*Fraxinus*) occur right across the northern hemisphere.[30] For example, the European ash (*F. excelsior*) and the narrow-leaved ash (*F. angustifolia*) are primarily found in Europe, whereas the Manchurian ash (*F. mandshurica*) is found in Asia and the black ash (*F. nigra*) is found in continental North America.

Today there are approximately 48 species of ash tree that occur across large parts of the northern hemisphere.[31] Their distribution is largely determined by the prevalence of low winter temperatures and therefore none can be found in Arctic regions. For similar reasons, ash trees are also generally found below altitudes of 1,000 m (3,280 ft).

North American Ashes

Of the 48, some nineteen species of ash naturally occur on the North American continent and in all but one case they belong to the three sections of the genus *Fraxinus* – namely *Dipetalae*, *Melioides* and *Pauciflorae*.

The section *Dipetalae* comprises three species that belong to the most ancient 'section' of *Fraxinus*, and are generally located in southern areas of the North American continent.[32] These are characterized by their peculiar rectangular, cross-sectioned twigs and branches, at odds with the cylindrical twigs and branches typical of other species of ash. *Fraxinus dipetela*, also known as the Californian or two-petal ash, is a species that takes the form of either a shrub or a small deciduous tree that rarely grows more than 7 m (23 ft) tall. It is native to the southwestern United States and some limited areas of northern Baja California in Mexico. It has very simple flowers that have a sweet scent and, like its earliest ancestor, this ancient species of *Fraxinus* is insect-pollinated.

Fraxinus anomala, also known as the single-leaf ash, is a species that has a restricted range in the southwestern United States, particularly in Nevada and Utah, with small populations in parts of northern Mexico. It takes the form of shrubs or small trees up to 6 m (20 ft) tall and is identifiable by its simple leaves, rather than the compound pinnate leaves more often associated with *Fraxinus* species. It is drought-tolerant and can live in a variety of dry habitats, including desert scrubland.

Fraxinus quadrangulata, also known as the blue ash, is native to a number of central U.S. states, including Illinois, Iowa, Missouri, Ohio and Kentucky. It takes the form of a medium-sized tree between 10 and 25 metres (32–80 ft) tall. Its specific name, *quadrangulata*, derives from the fact that its twigs have four corky ridges that create a rectangular cross-section to the branches and twigs. It has relatively large leaves and leaflets and small purple flowers. It is a commercial timber species in the U.S. and its common name derives from the blue dye that can be extracted from the inner bark.

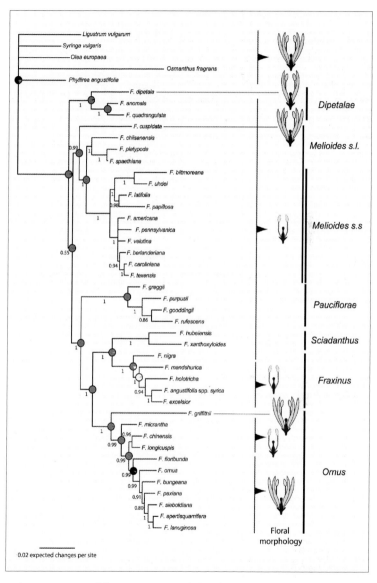

The main sections of the genus *Fraxinus* and the relative morphology of their flowers.

Ash trees play a central role in North American forests.

The section *Melioides* comprises twelve species of ash, and these occur more widely across the North American continent, from Canada to northern Mexico.[33] Ash species in this section include a number of major commercial timber trees, such as *F. americana*, *F. latifolia* and *F. pennsylvanica*.

F. americana, known as white ash or Biltmore ash, is the most common and commercially important type of ash tree in North America. *F. americana* is widely distributed across North America with populations occurring across most of the eastern half of the United States. The distribution also extends a short distance into eastern Canada and northern Mexico. It grows best on well-drained soils and forms a major component of a number of North American forest types, where it forms mixed forest stands with species such as northern red oak, white oak, sugar maple and eastern hemlock. White ash is relatively demanding in terms of both the moisture content and fertility of the substrate on which it grows. It is found from sea level to altitudes of up to 1,000 metres in areas such as the Cumberland Mountains.

The trunk of white ash trees is generally tall, straight and encased in dark grey bark, with a distinctive diamond-shaped ridge-and-furrow pattern. The trees are dioecious (male and female flowers on separate trees), with small green to purplish flowers. *F. americana* is deciduous and is a popular ornamental tree because of its colourful autumnal foliage, which ranges from yellow to red to bronze.

In addition to its commercial value, white ash is important to a range of wildlife. It is significant for its value as browse for large mammals such as deer and moose and also for the cover it provides. White ash trees produce prodigious numbers of winged seeds (samaras) each year that are eaten by a wide variety of wildlife, including foxes, squirrels, mice, wood ducks, cardinals, northern bobwhites, grouse and finches.[34] Moreover, the tendency for old trees to form holes and hollows means that nesting opportunities are created for a variety of animals such as woodpeckers, owls, nuthatches and grey squirrels.

Adult harvest mice include ash leaves and samara in their diets.

F. latifolia, or Oregon ash, is the only ash species native to the Pacific Northwest of the USA, ranging along the eastern edges of Washington and Oregon and extending into the Sierra Nevada in California. Oregon ash trees thrive in mild maritime climates and are adapted to live on heavy, often waterlogged, soils. They generally range from sea level to altitudes of up to 1,000 m (although occasional examples can be found above 1,500 m (4,920 ft) in the Sierra Nevada). When mature, the tree can reach between 10 and 20 m (32–65 ft) tall and 2 m in diameter. The large leaves can be up to 30 cm long (12 in.) and generally comprise seven to nine leaflets. Unusually the leaves tend to be more ovate than pinnate in shape.

F. pennsylvanica, known as the green or red ash, is North America's most widely distributed ash tree, occurring across much of the central and eastern USA and extending into Nova Scotia, Canada. It is a tree that stands 12–20 m (40–65 ft) tall with leaves comprising seven to nine large pinnate leaflets. It is a pioneer species that colonizes river banks and disturbed land. It has a large crop of 2–4-cm-long (0.7–1.5 in.) samara, which are an important food source to local wildlife. The green ash is widely used as an amenity tree in towns and cities across

Red-bellied woodpecker perched on tree limb.

North America. It is an important riverside tree, providing cool shade and an important habitat for riparian species, and its dense network of roots also helps stabilize riverbanks.

The section *Pauciflorae* comprises a further five species of ash: *F. dubia*, *F. gooddingii*, *F. greggii*, *F. purpusii* and *F. refescens*.[35] These ashes are all native to North America, but tend to occur in the arid regions of the southwestern USA, Mexico and as far south as Guatemala. They are generally small trees or shrubs with sparse clusters of 'naked' flowers, which are wind-pollinated. They are also distinguished by their small leathery leaves.

F. greggii and *gooddingii* are very similar but display subtle adaptations to the slightly differing ecological niches that they occupy. *F. greggii* is a small (2–4 m), multi-stemmed tree known as the little-leaved ash. Twigs are round in cross-section and there are one to five small leaflets that form a compound leaf. *F. greggii* is found on cliffs and bluffs and in canyon bottoms between altitudes of 400 and 1,400 m (1,310–4,590 ft) in western Texas and north-central Mexico.

F. gooddingii, which is closely related to *F. greggii*, is also a rare ash that typically grows as a dense shrub or small multi-stemmed tree, generally growing to a height of between 1.5 and 4 m (4–14 ft) tall. *F. gooddingii* has small leathery leaves with seven to eleven pinnate leaflets 1–2 cm long. It has a restricted range of parts of Arizona and Sonora, Mexico, with one population occurring on the remote island of Isla Tiburón in the Gulf of California.

The two species are generally distinguished by *F. gooddingii* having less hairy twigs and buds, more leaflets (five to nine instead of three to seven) and the occurrence of tiny white multi-radiate hairs on twigs, buds and petioles when young.[36] Otherwise these *Fraxinus* are often indistinguishable.

The final species found in North America is *F. nigra*, or the black ash. This belongs to the latest section of *Fraxinus* (in terms of evolution), which uncharacteristically has examples in all the continents of the northern hemisphere. *F. nigra* is a slow-growing tree found in the boreal woodlands of the eastern United States. The black ash

Mature Oregon ash, *F. latifolia.*

37

typically grows on poorly drained, waterlogged soils such as bogs and river margins. It can grow to heights in excess of 20 m (65 ft) and has a small open crown with distinctive ascending branches supporting seven to eleven oval or lance-shaped leaflets. It has small, not particularly noticeable flowers that emerge slightly in advance of the leaf burst in spring. The black ash has been used by Native Americans for a number of purposes, including basket making, and for this reason it is sometimes referred to as the basket or hoop ash. Like other ash species the seeds form an important part of the diet of small mammals and songbirds and the leaves and twigs are particularly palatable to deer and moose.[37]

Eurasian Ashes

There are approximately 24 species of *Fraxinus* in Asia and Europe today belonging to three sections: *Ornus*, which has sixteen species; *Sciadanthus*, which has just two examples; and *Fraxinus*, which has a further four.[38] (Note: not all *Fraxinus* species have been definitively placed in the six sections. Some are still awaiting being placed into their relevant section.)

Ornus is the largest section of the genus *Fraxinus*. Ash tree species in this section occur across Asia and Europe and are distinct from those of all other sections in their subtly different system of flowering. In *Ornus* species, the flowers are borne on shoots of the current year and emerge from terminal buds along with the new leaves. In all other sections the inflorescences emerge from lateral shoots from the previous year. Trees in the section *Ornus* are often referred to as 'flowering ashes' because of their showy flowers, and as such are often used as ornamentals.

F. griffithii is one of only two species of ash that is nearly evergreen. It grows to between 10 and 20 m tall (32–65 ft) and has large leaves (10–25 cm or 4–10 in., long) with five to seven leaflets. It occurs in

Green ash, *F. pennsylvanica*, in autumn colour.

The flowering ash *F. ornus* is a small decorative tree that can be found widely across the Mediterranean.

F. griffithii is a common street tree in Japan.

tropical Southeast Asia, typically on steep slopes, near villages and by rivers between altitudes of 1,000 and 2,000 m. Typical to the section *Ornus*, it has sweetly scented flowers. It occurs in the Chinese provinces of Fujian, Guangdong, Guangxi, Hainan, Hubei and Hunan, as well as in parts of Taiwan, Bangladesh, India, Indonesia, Myanmar, the Philippines, Vietnam and the Ryukyu islands of Japan.

F. chinensis comprises a range of taxa for which there is no agreed status. However, two broad subspecies are generally accepted: *F. chinensis* subspecies *chinensis*, which tends to be found in more southerly populations in countries such as China, Korea, Vietnam and Thailand; and *F. chinensis* subspecies *rhynchophylla*, which tends to have more northerly populations in China, Korea, Russia and Japan. The trees can grow to 3–20 m (10–65 ft) and have the distinctive compound leaf, generally comprising three to seven leaflets, which can vary quite widely in shape from ovate to lanceolate or even elliptic. They typically grow on sloped terrain, alongside rivers and roads, and in mixed woods between altitudes of 800 and 2,300 m

(2,625–7,545 ft).[39] Both have been part of the Chinese pharmaco-poeia for more than 2,000 years.

F. ornus has been called the 'manna ash' since at least the times of ancient Greece more than 2,000 years ago because it was, and still is, cultivated in order to collect a white exudate called 'manna', which contains a sweet component called mannitol that can be used as a sugar alternative. This is harvested by making incisions in the bark during the late summer. The manna ash is a small- to medium-sized

The developing ovules of *F. xanthoxyloides.*

tree (15–25 m, or 50–80 ft) that occurs primarily in southwestern Europe – from southern France to western Turkey, as well as on many of the Mediterranean islands. It often has an asymmetrical canopy and compound leaves comprising five to nine obovate, serrated leaflets. Its most distinctive feature is the large showy clusters of scented, creamy white flowers.

The section *Sciadanthus* consists of only two Old World species: *F. xanthoxyloides*, known as the Afghan or Algerian ash, which is found from Morocco and Algeria through the Middle East to the Himalayas, and *F. hubeiensis*, which is a threatened species endemic to Hubei province in China.[40] They both take the form of relatively small trees or shrubs. They have fairly small leaves and flowers, which are both polygamous and wind-pollinated.

F. xanthoxyloides takes the form of small deciduous shrubs and trees up to 7 m (23 ft) tall with smooth compound leaves which are typically 8–12 cm (3–5 in.) long. The flowers are polygamous and generally appear before the emergence of the leaves. They are found mainly in the drier temperate valleys such as the Sutlej Valley in Afghanistan. They are generally found between elevations of 1,000 and 2,800 m (3,280–9,190 ft) Their range includes Xizang region, China,[41] as well as Aghanistan, India, Kashmir, Pakistan and North Africa.

F. hubeiensis was first described in 1979.[42] It is closely related to the species *F. xanthoxyloides*, although it has a much more restricted range, being found almost entirely in Hubei province, China. It is a handsome tree that grows to heights of 19 m (62 ft) and has compound leaves 7–15 cm (3–6 in.) in length comprising seven to nine leaflets. As with much of the taxonomy of ash trees, *F. hubeiensis* is a curiosity, having more molecular similarities to *F. xanthoxyloides* than to its closer Chinese relatives.

The last section of *Fraxinus* to evolve is called *Fraxinus*. The species in this section comprise *F. nigra*, *F. excelsior*, *F. angustifolia*, *F. platypoda and F. mandshurica*. These species are not as geo-specific as other sections, with representatives in all the continents of the northern hemisphere.

For example, *F. nigra* is native to North America, while *F. mandshurica* is native to Northeast Asia and *F. angustifolia* is found in Europe. Despite occurring on two separate continents, *F. mandshurica* and *F. nigra* are so morphologically similar that they are often referred to as geographical races rather than distinct species.

F. mandshurica, commonly known as the Manchurian ash, is a medium to large deciduous tree that grows to 20–30 m (65–100 ft). It has a large distribution in Asia, covering some 23 degrees of latitude and 46 degrees of longitude in eastern Russia, northern China, Korea and Japan. It can tolerate a variety of soil types and is able to grow in both swampy areas and river valleys. It has compound leaves that can grow to 30–45 cm (12–18 in.), comprising seven to thirteen pinnate leaflets that have a distinctly toothed edge. The small, naked, greenish-yellow flowers are less showy than in other types of *Fraxinus*, particularly the section *Ornus*, and emerge in early spring in advance of the leaves.

F. angustifolia, also known as the narrow-leaved ash, is very similar in form to *F. excelsior*, growing to a height of around 25 m (80 ft) when mature. The five to thirteen leaflets that compose the compound leaf (15–25 cm, or 6–10 in., long) are distinctive as they are slightly narrower and smaller than those of *F. excelsior*. It thrives on riverbanks and flood plains, and is an important component of rare riverplain woodland. It generally has a more southerly distribution than *F. excelsior*, occurring in central and southern Europe, North Africa and east through Turkey and Iran to the western fringes of Russia. There is much morphological variation in *F. angustifolia* and it has a tendency to hybridize with its closest relative, *F. excelsior*, creating complications in identification.

F. excelsior (Common or European Ash): A Case Study

The European or common ash, *F. excelsior*, is one of the most commercially important and common tree species in Europe. For this reason, it is probably the species of ash for which most research exists and, as such, seems to be the most logical species to look at as an insight into the *Fraxinus* species generally.

It is a native tree that is found in forests, hedgerows and towns and cities across most of Europe. European ash, in favourable conditions, generally lives to around two hundred years old. However, when managed by coppicing or pollarding, the trees can often live to much greater ages, sometimes exceeding three hundred years old.[43] In the vast, largely untouched forests of Białowieża National Park in Poland, it is not uncommon to find ash trees aged 250–400 years.[44]

The European ash is widely distributed, found from Ireland and northern Spain in the west to the course of the Volga river in Russia

A giant European ash tree, *F. excelsior*, several hundred years old, standing in a Devon meadow.

in the east.[45] The European ash's most northern population is found in Norway at 64 degrees north, while populations which extend south reach into the mountainous areas of the Mediterranean, through northern parts of Spain, Italy and Greece up to altitudes of 1,600 m (1 mi.) (some trees have been recorded as living as high as 2,500 m in parts of Iran). The ash tree's northern limit closely follows the January zero-degrees-centigrade isotherm across Northern Europe.[46] This is because of the tree's susceptibility to late spring frosts. However, ash trees growing at high altitude in central and southern Europe have developed an interesting adaptation: they have become more tolerant to the intensely cold winters, but require a longer over-all growing season by way of compensation. In general, the extent of the European ash populations in western Europe is largely determined

The cluster of two petalled flowers of *F. excelsior*.

The range of *Fraxinus* species is limited by the zero-degrees winter isotherm in the northern hemisphere.

by the tree's inability to withstand cold winters (and in particular late spring frosts) and hot, dry summers.

Ash trees cover around 2.6 per cent of the land surface and account for 2.7 per cent of the timber removed annually from European forests.[47] Interestingly, the population of European ash has increased markedly over the last fifty years, which is attributed to the change in human habitation from rural communities to more urban ones.

The European ash usually occurs in groups within mixed broad-leaved woodland or as pure stands. It can also form associations with a wide range of trees in a wide range of environmental conditions. To a lesser extent ash trees can be found as scattered trees across the rural and urban environments. The European ash is also a major component of hedgerows. In Great Britain ash was recorded as the most common 'standard' hedgerow tree species, accounting for an estimated combined length of hedge of 98,900 km (61,450 mi.).[48] Based on an ash tree being located every 8 metres, this would give an estimated hedgerow population of nearly 20 million trees in the UK.[49]

The European ash is the tallest and largest in volume of any of the *Fraxinus* genus, with examples reaching more than 40 m (130 ft)

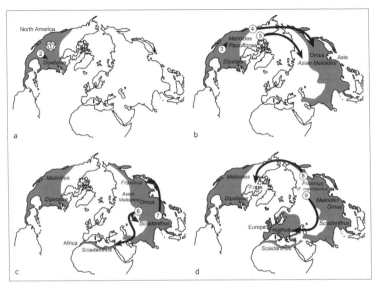

Map illustrating the major events in the biogeographic history of the genus *Fraxinus* and its eventual expansion into Europe.

in height and over 8 m (26 ft) in girth. The largest recorded European ash is located in Portugal and has a girth of around 13 m (43 ft),[50] and trees growing to heights in excess of 45 m (148 ft) have been reported from an exceptional stand in the West Carpathians in Eastern Europe.[51]

F. excelsior is an ecologically flexible species and can be both a pioneer and a permanent component of a forest. For example, the species is particularly shade-tolerant for its first seven years or so, up to about 4 m (13 ft) tall. This is part of its competitive colonizing strategy where it has the ability to establish a permanent carpet of seedlings under relatively shady conditions, which are then poised to take advantage of any gaps that may form in the canopy. However, once a fully grown tree, European ash becomes less shade-tolerant and relatively light-demanding. As a result, *F. excelsior* is a highly competitive forest tree species and one that regenerates well in partially shaded or open conditions.

The European ash can grow on many soil types and thrive under a wide range of pH levels. However, European ash grows best on fertile soils,[52] and usually occurs in woodlands formed on calcareous soil, which are not too acidic, and where the water table (the depth at which there is permanently water in the soil) is between 40 and 100 cm (16–40 in.) below the surface. European ash can also become the dominant tree on lowland clay soil sites that are prone to flooding, as well as on relatively dry calcareous screes and fertile alluvial soils. European ash often forms part of a variety of broadleaved forest communities, including fertile beech forest, oak/ash/beech/alder forest and, in areas of inundation, ash/alder swamp forest.

Research into *F. excelsior* in Germany has identified at least two forms of the European ash, each related to the type of soil on which they grow. Those associated with moist, alluvial soils are referred to as 'water ash', whereas trees that grow on the drier, thinner, calcareous soils are referred to as 'limestone ash'. Close examination of these two subspecies reveals that there are distinct morphological differences, even in the properties of their timber. However, these types of ash are simply considered the two ends of a continuum of morphological variation determined by local environmental factors. This highlights just one of the difficulties in confidently identifying *F. excelsior*. Unhelpfully, ash trees are not neatly homogeneous in their morphology; instead they display a range of subtle variations of leaves, seeds and flowers that can present as a number of subspecies or races. They can also hybridize with other ash species, particularly the narrow-leaved ash *F. angustifolia*. *F. excelsior* is closely related to *F. angustifolia*, and the two are often confused. However, of all the characteristics that distinguish the two species, it is the type of inflorescence (flowers) that is the most reliable. *F. excelsior* has a branched inflorescence (compound panicle), whereas *F. angustifolia* an unbranched simple raceme or stem.

The leaves of *F. excelsior* display the typical morphology of ash trees, with large compound leaves measuring between 20 and 35 cm (8–14 in.), comprising seven to fifteen oddly pinnate leaflets. The leaflets are glabrous (hairless), except for tufts of hair at the base of the underside of the midrib, and have a slightly asymmetrical base and serrated margins. The leaves appear relatively late in spring, sometimes as late as June, and fall again in the early autumn between late October and the first half of November.[53] It appears that the onset of autumn leaf fall does not change with altitude, therefore it is believed that the leaf fall is likely to be controlled by day length rather than by temperature. The winter leaf buds are black and felted.

Flowers

The mating system of European ash is a continuum from pure male (producing pollen) to pure female individuals (producing seeds), with differing degrees of hermaphroditism in between.[54] It appears, however, that ash might be what is known as sub-dioecious or functionally dioecious.[55] This is where some hermaphrodite flower clusters are functionally female in that they produce seeds.

The European ash is a wind-pollinated and wind-dispersed species. The simple apetalous flowers (in which the calyx and corona are absent) are borne on axillary panicles, each of which consists of a single ovary containing four ovules. The individual inflorescence buds tend to open between March and April. However, in general the flowers open in a specific order, with flowers of the male tree opening first, followed by hermaphroditic flowers and finally the female flowers.[56] In studies carried out in Northern Ireland it was found that of the total flowers produced more than half were male but just 1.4 per cent were female. However, many of the flowers of hermaphroditic trees acted as functionally female, bringing up the proportion of functionally female flowers overall to 23.3 per cent. There were differences between the trees in other ways, too. Male trees tended to produce a similar number of flowers each year whereas the total number of flowers on female trees varied significantly from year to year.[57]

European Ash trees generally begin to produce flowers for the first time at around fifteen to twenty years old when growing as isolated trees, but in a closed-canopy forest trees can be thirty years old before they flower for the first time. F. excelsior has been estimated to produce between 4 million and 5 million flowers per tree each year.[58]

The pollen is microscopic at just 22.5 micrometres (0.0009 in.) in diameter,[59] and is therefore well adapted to wind pollination. However, various insects are known to frequent ash flowers, which are an important source of pollen for insects such as honeybees (*Apis mellifera* L.). Birds such as blue tits (*Cyanistes caeruleus*) can also disperse pollen. Blue tits have been observed visiting ash flowers and when subsequently caught in mist nets were shown to have an average of 75 grains per centimetre on their heads – enough to make them minor pollinators, too.[60]

Samara

Generally, just one of the four ovules develops into a papery winged seed (samara), which is about 3–4 cm long and pale brown when ripe. These can persist in clusters on the tree until well into the winter before being dispersed by wind. Open-grown European ash trees begin to produce winged seeds between twenty and 25 years old. However, ash trees growing in closed forest can take as long as thirty to forty years to get to samara-producing maturity, after which the trees will continue producing flowers and samara until the demise of the tree, generally between 150 and 220 years old.[61]

The samaras are fully grown by the beginning of July. However, the enclosed seeds do not reach their full length until the beginning of August. The first viable seeds begin to fall from ash trees in September, continuing throughout the winter until March.[62] Once on the ground they remain dormant for at least two winters, finally germinating in late April or early May.[63] Ash has been recorded as producing up to 140,000 seeds per tree annually (around 10 kg, or 22 lb).

Seeds contained in samaras are mainly dispersed on the wind, but seeds can also disperse along watercourses, and survive in water for several weeks.[64] Fruit fall is distributed throughout the winter and spring, but fruit production varies from year to year. This variation is balanced by seed viability lasting for two to three years.[65] The embryo inside ripe seeds is only partly developed at the time of

An ancient ash tree with hollows, crevices and deadwood in the crown provides a wide array of niches for invertebrates, lichens and ferns.

dispersal and the seeds remain dormant for at least one vegetation period (sometimes for two or more) before germinating.[66]

Trunk, Branches and Bark

European ash trunks are what is known as orthotropic (straight-growing) and monopodial, meaning that they prioritize vertical growth rather than growth in girth, especially when young. They also self-prune (the lower branches drop off) to leave clean, straight trunks, which is why they are of considerable commercial interest. The trunk is smooth and pale grey when young; however, as the tree gets older, the bark develops a fissured latticework of diamond-shaped furrows that eventually become encrusted with mosses, ferns and lichens.

The growth in the diameter of the trunk has been demonstrated to be more closely linked to rainfall than to temperature.[67] For example, annual ring widths in European ash have been seen to be positively correlated with precipitation and air temperature of the previous winter (December to February), rather than with precipitation in spring (May and June).[68] Research in Germany has highlighted that of all the main commercial trees studied, the radial growth of the trunk in a growing season is highest in ash.[69] Mature ash trees can grow radially at twice the rate of beech (*Fagus sylvatica*), for example, and nearly four times the rate of oak (*Quercus sp.*) – circa 5.8 mm (0.2 in.) compared to around 2.8 mm (0.1 in.) and 1.5 mm (0.05 in.) per year respectively.[70] The growth rate was highest in May for the European ash. The European ash is also unusual because of the nineteen broadleaved and conifer species investigated in the study the ash had the highest concentration of chlorophyll per unit area of bark (500 mg/m²) of any European tree.[71] The bark of *F. excelsior* is also rich in a number of complex chemical compounds, some of which have major medicinal potential. For example, the bark of a typical European ash comprises 32.5 per cent cellulose and 20 per cent hemi-cellulose and 25.8 per cent hydrophilic extractives, such as hydroxy coumarins and phenylethanoids, but very little in the way of tannins. For example hydroxy coumarins can be used by major pharmaceutical companies for the production of drugs similar to warfarin, a blood thinner.[72]

The timber of *Fraxinus* is known as ring-porous. This is because the vessels that develop in the early-season timber (which transport nutrients and water within the tree) are often large. Large enough to be seen with the naked eye in some cases, ranging in size between 80 and 170 micrometres (0.003–0.006 in.) in diameter. As the season progresses the tree lays down denser sapwood (cambium) with much smaller vessels (xylem and phloem) ranging from 10 to 70 micrometres (0.0004–0.002 in.).[73] In cross-section the annual rings are clearly visible in ash, with the two halves of the growing season often represented by slightly different bands of colour. The internal hydraulic architecture of *F. excelsior* has a number of unusual adaptations which help to ensure that the leaves in the canopy get the same access to water as those lower down.[74] In the European ash the water resistance in older and younger branches is determined differently. In lower, older branches, around 90 per cent of the water

An open canopy of mature ash trees lets light into the forest floor.

resistance comes from within the branch, whereas with newer and younger higher branches the same percentage of water resistance comes from within the leaf. It has been found that, to compensate, the leaves of two-year-old branches are served with much larger rachis (stems) with larger internal vessel diameters than older branches to help keep the access to water similar right throughout the canopy.

The density of European ash timber can vary quite widely from 551 to 790 kilograms per cubic metre.[75] Most of the differences in densities between timbers are caused by the speed of growth in European ash. As with most ring-porous timbers, the faster the rate of growth the larger the proportion of latewood in each annual ring, and therefore the higher the density. Conversely, the faster the ash grows the less strong the timber is along the grain, because of the thinner cell walls in the earlywood, and the easier it becomes to split.[76]

The branches of European ash are distinctive and help the observer to readily identify the ash tree from many metres away, particularly in its leafless state during the winter months. The superstructure of the typical ash tree comprises branches that subdivide in a manner not unlike a candelabra; the final twigs swoop down and then up like the skeletal fingers of an upturned hand, leading to them being popularly described as the 'witch's claw'.

Root System

The development of the roots of *F. excelsior* is known to vary widely according to the conditions in which the seedling or young tree finds itself. In a mixed stand of mainly deciduous trees, for example, the young ash roots tend to penetrate deeply, often to a depth of more than 2 m (6½ ft). However, as the tree matures, and the gap in the canopy closes, the roots begin to head towards the surface where they go on to develop a network of extremely fine roots less than 1 mm (0.04 in.) in diameter. These roots are at their greatest density in the top 5 cm (2 in.) of the soil.[77] However, this shallow, dense, rope-like

Ash trees have shallow, tenacious roots, allowing them to grow on very steep slopes.

root system is known to develop straight away in seedlings that are in open-grown situations, whereas the root plates of seedlings developing in woodland situations are quite different, having a smaller network of shallow roots. This indicates that the structure of the root plate is largely determined by access to light.[78] The growth of the root system has been recorded to continue vigorously throughout the ash tree's entire life. This means that an old ash tree can have a root system that is significantly greater in volume and biomass than that of beech (*Fagus sylvatica*), hornbeam (*Carpinus betulus* L.) or maple (*Acer sp.*) of similar trunk dimensions.[79] As with the majority of trees, growth of ash roots takes place whenever the soil temperature is above 4–6 °C (40–43 °F) below 30 °C (86 °F). The dense network of roots also enables trees to grow on very steep slopes with gradients of up to 80 per cent (72 degrees).[80]

Biodiversity

F. excelsior is one of Europe's most common and wide-ranging trees. It is common in forest, woodland and wood pasture, and also as individual free-standing trees in many rural areas. In addition, it is a common street tree and often forms a major component of hedges. While it does not live as long as an oak tree it still provides an important habitat, especially for epiphytes (species that simply live on the tree without being parasitic), owing to the high pH of the bark. For example, a survey conducted in northern Norway found 84 species of lichens, 72 species of bryophytes (mosses and ferns) and seventeen species of vascular plants living on the ash trees surveyed.[81] Even in areas such as Central London, ash trees have been recorded having as many as 74 lichens, fourteen mosses and seven fungal and three algal species on a single tree.[82] These figures highlight the fact that the European ash can support a similar number of total species to that of oak and may in fact have a richer bryophyte and lichen flora because of its alkaline bark.[83]

Much work in the United Kingdom into the associated biodiversity of *F. excelsior* has been undertaken because of the impending threat of the disease *Chalara fraxinea* (ash dieback). For example, the Joint Nature Conservation Committee (JNCC), an organization that advises the UK government on nationwide and international nature conservation issues, has compiled an exhaustive list of the species that depend on *F. excelsior* in one way or another in the UK. Many lichens

The goldfinch and other birds often eat ash seeds during hard winters.

Large ancient ash tree in the Lake District.

and bryophytes occur on the alkaline bark of the ash while thirteen species of birds are known to have an association with the ash tree, including the greater and lesser woodpeckers (*Dendrocopos major* and *minor*) and bullfinches (*Pyrrhula pyrrhula*), which can take large numbers of seeds during harsh winters. Lichens account for most of the biodiversity associated with ash with more than five hundred known species, while invertebrates account for around a further 240 species. So far, this research has identified a total of 1,058 species associated with British ash trees.[84]

These figures, however, tell only part of the story. For example, of the 547 species of lichen found on ash in Britain, three are critically endangered, nine endangered, twenty vulnerable and 52 near-threatened species.[85] JNCC's research has identified a total of around a hundred species that are either highly or uniquely (obligate) dependent on *Fraxinus* for their well-being. The majority of the other associated species (numbering more than nine hundred in Great Britain) appear to be able to thrive on other types of tree if necessary. However, there are parts of Europe where the potential loss of ash trees could have an even more devastating effect on biodiversity. A prominent example of this is found in Scandinavia. More than a third of the 174 species of lichens found on the Swedish island of Gotland could be lost if the ash trees are eradicated.[86] A number of these species have already moved from elm to ash in the wake of Dutch elm disease because of the similarity of its alkaline bark but with ash dieback there will be fewer suitable host trees to which they can transfer.

Ancient ash tree in Swaledale, North Yorkshire.

two

The Threatened Ash

Ash trees around the world today are dying at an unprecedented rate, in what could eventually be cited as one of the great tree extinctions of human times. One of the biggest threats to ash trees in particular, and global biodiversity in general, is the accidental introduction of non-indigenous, invasive pests and pathogens. In North America, for example, tens of millions of ash trees have already been lost to damage caused by a recently introduced beetle, known as the emerald ash borer (*Agrilus planipennis*). At the same time, in Europe, millions of trees have died as the result of the unchecked spread of the fungal infection ash dieback. These are just two examples of biotic factors seriously affecting ash tree numbers worldwide, which are in turn compounded by the abiotic pressures already faced by ash trees and their associated biodiversity, such as loss and fragmentation of forests, urbanization, pollution and climate change.

Ash Dieback

European ash (*F. excelsior*) is one of the most economically and ecologically significant forest trees in Europe. It occurs throughout much of the continent and its numbers can be measured in hundreds of millions, making it one of the most common deciduous tree species on the continent. However, the emergence of a new, invasive fungal disease, commonly known as ash dieback or *Chalara* ash dieback,

Infected ash trees suffering from ash dieback.

now threatens its existence.[1] *Chalara* ash dieback is a disease that is potentially lethal to European ash trees, threatening not only the trees but the many organisms that rely on the them for their survival. It is estimated that much of ash cover an area of around 2 million square kilometres (770,000 sq mi.), from Scandinavia to the mountains bordering the Mediterranean, which is already either infected or threatened by infection.[2]

Dieback in ash is a phenomenon that has been concerning scientists since it was first observed in British hedgerows in the 1960s.[3] At that time the likely causes were considered to be the dark ash bud moth (*Prays ruficeps*) along with the effects of abiotic factors such as periodic drought, late frosts and disturbance of roots by ditching.[4] However, a new causal agent of the latest outbreak of ash dieback, which first appeared in Poland in the 1990s, was identified in 2010 as being caused by a fungus formerly restricted to parts of Asia.

The outbreak rapidly spread beyond the national borders of Poland into neighbouring countries (in order of being reported: Lithuania, Latvia, Sweden, Germany, the Czech Republic, Denmark, Slovakia, Belarus, Estonia and Austria),[5] and now, some 25 years later, ash dieback has spread to almost all corners of Europe, including Britain, where it was first identified in a consignment of imported trees in 2012. The causal agent of the devastating outbreak of ash dieback was first described as the new species *Chalara fraxinea*;[6] however, this proved to be the asexual (anamorphic) stage of just one of two aggregated fungi that caused ash dieback.[7] The accepted name is a combination of the two distinct fungal species: *Hymenoscyphus fraxineus*.

Prays ruficeps (dark ash bud moth) is one of a number of species that are specifically dependent on ash trees as part of their life cycle.

The distinctive burnt-looking leaves of ash trees infected with ash dieback.

Ash dieback is transmitted by the release of *Hymenoscyphus frax-*
ineus ascospores, which are carried on the wind before alighting on
the leaves of ash trees.[8] The infection enters the ash through the
living leaves and stalks, after which the mycelium (fungal filaments)
develop intracellularly, passing between cells rather than through the
actual cell walls. It then invades the phloem, xylem and pith via the
ray parenchyma cells that transport materials laterally in the woody
tissue, and from there spreads throughout the tree.[9]

Once the fungus becomes established it progresses through the
entire tree (systemically) starting with the leaves, then the branches,
followed by the trunk and eventually causes the demise of the entire
tree. It has been observed to have lethal effects on trees in all stages of
development whether seedlings or mature trees. The first symptoms
of ash dieback are the noticeable leaf spots, wilting and the red-brown
discolouration of the leaves, petioles and rachis, and the presence of
bark cankers. Infected trees also develop lesions and necroses on
shoots, branches and stems.[10] This leads to a progressive dieback in
the crown. The fungus eventually kills its host by girdling the trunk,
for which there is no treatment.

The prognosis for ash trees is severe, with the predicted mortality
rate for *F. excelsior* across Europe believed to be around 95 per cent.[11]
This is based on what has already been observed in countries such as
Denmark and Sweden. In assessments of the likely impact of ash
dieback in countries such as Britain, the worst-case scenario is also
taken as 95 per cent mortality for the European ash. The enormity
of the threat to rural woodland and urban ash trees in European
countries can be understood when the sheer number of ash trees in
each country is known. In a recent report by the Tree Council, for
the UK government's Department for Environment, Food and Rural
Affairs (DEFRA), the populations of ash trees in the UK were estimated: between 27.2 and 60 million ash trees in non-woodland
situations (with a diameter greater than 4 cm at breast height); as
many as 125.9 million in woodlands with an area larger than 0.5 hectare; and potentially 1.75 billion saplings and seedlings in woodlands

and non-woodland situations.[12] It can be seen that the severity of the impact of ash dieback in Britain is likely to be far greater than the loss of 30 million elm trees in the 1980s caused by Dutch elm disease. In Sweden, *F. excelsior* has been declared to be under threat and in 2015 was declared endangered by the IUCN Red List.[13]

The predicted death of such a large number of European (*F. excelsior*) and narrow-leaved ash (*F. angustifolia*) will have a major effect on European ecosystems, significantly reducing tree biomass and in turn the environmental services they provide, such as local temperature regulation, pollution reduction and carbon sequestration. The loss will also have major consequences for biodiversity in European countries. Ash trees occur in a wide variety of locations, but ash dieback is likely to have the most severe effect on European biodiversity in the relatively rare floodplain ecosystems.[14] Here ash trees are a keystone species and their demise threatens the survival of a number of already vulnerable associated species. The loss of ash trees from roadsides, hedgerows, field margins and urban landscapes will also contribute to habitat and

A forest of dead ash trees in southern England.

Rare and vulnerable lichens such as lungwort rely on the alkaline bark
of old ash trees to thrive.

species loss. For example, ash is a key species for land snail diversity,[15] as well as for threatened bryophytes (ferns) and lichens.[16] Of particular concern is the potential loss of ancient and veteran ash trees, but even young *F. excelsior* trees are important for some threatened species, such as the butterfly the scarce fritillary (*Euphydryas maturna*), whose caterpillars rely on ash leaves as their main food.[17]

Research in the Netherlands has offered some hope with regard to developing a resistant strain of European ash tree with which to potentially repopulate Europe. In Utrecht researchers from Terra Nostra, a knowledge centre dedicated to the study of trees and soil, have estimated that the total number of trees in the city is around 160,000, of which just over 21,000 are ash trees. Of these the majority are *F. excelsior* or its cultivars. It was shown that cultivars of the same ash species can have different infection rates. For example, 100 per cent of *F. excelsior* (*Pendula*) and 90 per cent of *F. excelsior* (*Jaspidea*) were recorded as infected with ash dieback. In contrast only 2.8 per cent

Ash varieties such as *F. excelsior aurea* have shown some resistance to ash dieback.

of the *F. excelsior* (*Atlas*) were infected. *F. americana* also demonstrated some potential resistance, with less than 10 per cent of the city's trees becoming infected.[18]

Emerald Tree Borer

Ash trees in North America, and more recently in Europe, have been suffering from another major agent of destruction, this time in the form of the rather beautiful-looking beetle the emerald ash borer

(EAB), *Agrilus planipennis*. The emerald ash borer is a beetle native to China, Japan, Korea, Mongolia and the far east of Russia, where it feeds primarily on species such as *F. chinensis* and *F. mandshurica* and tends to exist largely in balance with the local ecosystems.

This large, attractive, metallic green beetle was first discovered in the USA, far beyond its normal range, in 2002, where it was thought to have entered the country in packing crates carrying items from Southeast Asia.[19] Since then, in the absence of natural predators and with an abundant food source provided by millions of ash trees of various species in the U.S., the EAB population has exploded. Within a decade, the EAB had expanded across fourteen U.S. states (Illinois, Indiana, Kentucky, Maryland, Michigan, Minnesota, Missouri, New York, Ohio, Pennsylvania, Tennessee, Virginia, West Virginia and Wisconsin) as well as two Canadian provinces (Ontario and Quebec), covering an area of some 13,000 square kilometres (5,020 sq. mi.) and devastating local ash tree populations.[20] Today it is believed to be present in 25 states and provinces in the U.S. and Canada, and

Red deer (*Cervus elaphus*) fawn eating ash tree leaf.

The accidental introduction of the emerald ash borer (*Agrilus planipennis*) has caused the death of millions of ash trees in North America.

responsible for the death of around 50 to 100 million ash trees.[21] Each year both the population and range of EAB continue to increase, creating a bleak outlook for the estimated billions of North American ash trees. The U.S. Forest Department estimates that before the arrival of the emerald ash borer there were around 8 billion ash trees of commercial size in the U.S., with a value of $282.25 billion.[22] 'The cost to the U.S. economy over the ten-year period from 2009 to 2019 – in terms of tree removal and replacement alone – is expected to exceed $10bn.'[23]

The adult emerald ash borer beetles appear from mid-May to late July and feed almost exclusively on ash foliage. The adult females live for about 22 days and males slightly less. They are slim, elongated iridescent blue-green beetles between 7.5 and 13.5 mm long and around 1–3.5 mm (0.04–0.14 in.) wide. The narrow body is somewhat wedge-shaped and both the head and the vertex are shield-shaped. The compound eyes are distinctly kidney-shaped and

New England forest in autumn containing *F. americana* and other ash species.

bronze-coloured. The adult beetles feed on the leaves of *Fraxinus* sp. but cause very little damage during their short lives. After fifteen or twenty days of feeding on the ash leaves the females begin to lay between thirty and sixty eggs on host trees such as white ash (*F. americana*), green ash (*F. pennsylvanica*) and black ash (*F. nigra*). They generally lay a few at a time, concealing them under flaps of bark or within crevices. Each egg is oval in shape and approximately 1 mm long.

It is with the emergence of the creamy white larvae that serious damage to the trees begins to occur. The larvae feed voraciously as they burrow through the upper layers of the wood of the ash tree for several weeks, creating large S-shaped galleries beneath the bark. Each larva can reach a size of up to 30 mm (1 in.) long and can create tunnels between 20 and 30 cm (8–12 in.) in length.[24] As they feed they damage the essential xylem and phloem vessels of the live wood that carry water and nutrients throughout the tree. It is the damage to these vital vessels that causes the demise of the tree. These larval galleries can also become filled with brown frass (excrement and

It is the damage caused by the burrowing emerald ash borer larvae that leads to the eventual death of ash trees.

other debris left by the burrowing insect) and can, in some cases, become large enough to girdle entire branches or even the trunk, which kills the tree.

The symptoms of EAB generally only become recognizable once the internal damage to the tree has already become severe. Trees start by exhibiting a yellowing of the canopy and some dieback in the crown. Excessive epicormic growth (shoots growing directly out of the trunk) and longitudinal splits in the trunk are also key indicators of EAB infestation. As pieces of bark become detached from the stem the extensive snaking galleries created by the larvae become visible. Generally, a small tree will die within a year but a mature tree can take as long as four years to succumb. There are also tell-tale D-shaped holes in the stem where large beetles have exited after pupating.

Studies have revealed that adult EAB beetles are good fliers and amazingly adept at finding and colonizing ash trees. The EAB uses both vision and the detection of the chemicals released by ash leaves and bark. To do this they have chemosensors which are very sensitive to certain chemicals in the air. They are particularly good at detecting the distinctive chemical signature of a damaged or stressed ash tree.[25] Curiously most EAB will range for less than a kilometre; however, some individuals travel greater distances of up to 10 km (6 mi.) looking for host trees.

Various species of ash trees have been extensively planted in urban landscapes and along roadsides in the U.S. for decades. These trees, which number in their hundreds of millions, are now threatened by the EAB. This has not only a major environmental and ecological impact, but a substantial cost implication.

The long-term ramifications of ash mortality in forests and riparian (river edge) settings are not yet known, but can be expected to cascade through ecosystems. Nutrient cycling, hydrology, composition of herbaceous plants and the habitats available for birds, mammals, insects and other animals are all likely to be affected.

One bright spot in the EAB saga involves the blue ash (*F. quadrangulata*). U.S. scientists have recently determined that blue ash is relatively resistant to EAB, making it likely that this will be the species that best survives the invasion.[26] Understanding more about the chemical and physical traits that underlie blue ash resistance may eventually lead to selective propagation of resistant ash cultivars.

The EAB beetle was also discovered close to Moscow in 2003, and is suspected to be present in Sweden, from where it could invade the rest of Europe.[27] Given the enormous impact it has had in North America, the beetle is now on the alert list of many global, regional and national plant protection and environment protection organizations worldwide. Recently a team of two UK and two Russian scientists found that the emerald ash borer population had spread 235 km (146 mi.) west of Moscow and 220 km (137 mi.) south of the Russian capital city.[28] The EAB is a major threat to the European ash (*F. excelsior*) and should it become established in the ash-dominated forests south of Moscow, the sheer abundance and almost continuous presence of ash trees across this wide area means that EAB has the opportunity to invade a number of other European countries on a broad front. The only potential comfort is that the European species of ash are more closely related in their evolution to those of Asia than are their North American cousins, which may mean that they possess genetically inherited defences against the EAB.

Other Diseases and Abiotic Causes of Ash Decline

Ash decline is the rapid decline of ash trees leading to sparse crown growth and dieback. It may be caused by many factors, including environmental concerns like drought, poor soil conditions or root damage, or biotic factors such as ash-specific pathogens.

Ash yellows is a disease caused by *Candidatus fraxinii*, which affects the tree's vascular system or its phloem sieve tubes.[29] It is presently only found in North America. The ash yellows disease cycle is still a mystery. Experts think it is transmitted by insect vectors, such as leafhoppers, which also transmit viruses. Symptoms seem to appear more often on white ash (*F. americana*) than green ash (*F. pennsylvanica*).

However, it can also affect at least ten other ashes, including black ash (*F. nigra*) and blue ash (*F. quadrangulata*). Early signs of ash yellows phytoplasma infection include the shortening of the internodes on twigs, the tufted appearance of the leaves (often referred to as witch's broom), a thinning crown and premature autumn colour. Apart from molecular testing, using polymerase chain reaction, the most reliable symptom by which to identify yellows is the presence of witch's broom.

There are a variety of other organisms that can also have a negative effect on the health of ash tree species, such as the ash weevil (*Stereonychus fraxini*), the large pine weevil (*Hylobius abietis* L.), the European red click beetle (*Elater ferrugineus* L.) and the saproxylic beetle (*Prionychus ater*). Ash trees are also particularly susceptible to infection by several types of fungus (*Phytophthora spp.*).

Abiotic Factors

Ash trees are also susceptible, like many other tree species, to abiotic factors such as climate change, forest fragmentation, air and water pollution, and physical stress caused by compaction around the

An ash tree almost cut down by a beaver.

root plate caused by agricultural and other types of vehicles. Modern farming practices such as deep ploughing and ditching can also cause stress to ash trees by damaging the root system, which is typically close to the surface.

In many ways ash trees are less susceptible to climate change than many other tree species. One of their range-limiting factors is the northern zero-degree isotherm because of their susceptibility to late frost. However, global warming may well increase the range of *Fraxinus* northwards into Scandinavia and Siberia. In addition, the length of time required for seeds to germinate when exposed to low temperatures over winter is relatively short, therefore ash seedlings will still be able to germinate even with rising winter temperatures.[30] Ash are also relatively drought-resistant and therefore reasonably well able to withstand the predicted droughts associated with climate change. Ash responds well to small increases in the proportion of CO_2 in the atmosphere but higher levels inhibit growth.[31]

Shoots, bark, flowers and fruits of ash are utilized by foraging herbivores. However, browsing may harm ash and even reduce population sizes, affecting the reproduction success of the species.[32] High grazing intensity can prevent ash seedlings becoming established, and even the grazing of ash in different seasons of the year can also have an important impact on the species. For example, more damage is caused to young ash trees by grazers in autumn than in spring.[33]

The European beaver (*Castor fiber* L.) can cause some damage to ash trees. In the Czech Republic, their diet includes fourteen tree species with tree diameters ranging up to 2 m (6½ ft): the preferred species were ash, with a diameter of 1–10 cm (0.4–4 in.), and willow (*Salix* spp.), with a diameter of 11–20 cm (4.3–8 in.).[34] Although ash regenerates easily, damage by browsers can locally reduce the amount of valuable timber.

three
The Mythology of Ash

Trees appear in the mythologies of cultures throughout history and across the globe, and while there are many local variations, there are also a number of common themes that have helped to explain creation and the structure of the universe to people via the metaphor of a great tree. According to quantum physics, the 'Big Bang' marked the beginning of the universe. However, in many origin myths the same event is represented by a point of existence coming into being: an absolute beginning, which first extends up and down into a vertical line, becoming the axis of the world (*axis mundi*), like the trunk of a tree. From this, lateral branches develop, creating the basic structure of the 'World Tree', 'Cosmic Tree' or 'Tree of Life'.

According to Gary R. Varner in his book *The Mythic Forest: The Green Man and the Spirit of Nature*, 'The "cosmic tree" is a universal symbol. Legends of existence appear in Native American, Scandinavian, Mesoamerican, Siberian, Asian, African, Middle Eastern and Indian lore. It's perhaps one of the oldest universal tales of mankind.'[1] The concept of the 'World Tree' or 'Cosmic Tree' is that it contains everything visible and invisible in the universe. As it continues to branch and root it provides the infinitely complex skeleton or superstructure upon which the whole of existence is supported. The value of using a giant tree as a metaphor for the structure of the universe is that it allows for the interconnectedness of everything. As a concept, it means that everyone and everything are part of the same and that

the actions of humans, just as much as those of gods and demons, have an effect on the universe. It also appears to refer, via metaphor, to the innate and instinctive understanding that humanoids spent the majority of their evolution among trees and in forests.

The World Tree explains the universe as an organic, interconnected entity that, in its simplest form, comprises three levels – the Underworld, Middle World and Upper Realms. The Middle World is where humans reside, the Upper Realms – the branches and canopy – are where the deities and the divine exist, while the Underworld is where various monsters and hells are said to be located.

The above description of the World Tree is a simplified and generic account, but it gives us a sense of this seemingly pan-global belief system that explains the complex idea of existence via the easily visualized concept of a huge tree. On a local level, the type of tree chosen to represent the World Tree or 'Tree of Life' varies from location to location, but it is generally one that has been fundamental to the development or survival of a particular culture. For example, in North America a number of native peoples revere the cottonwood (*Populus sect. Aigeiros*) tree as the Tree of Life,[2] while in Central America some indigenous groups consider the towering kapok tree (*Ceiba pentandra*) to be representative of the Tree of Life.[3] Similarly, for peoples of the Indian subcontinent it was the pipal fig (*Ficus religiosa*) that took on the role of the World Tree: it was under a pipal fig that Buddha was said to have gained enlightenment. The symbol of three cosmic regions connected by a tree is common to Norse, Vedic Indian and Chinese mythologies.[4]

Historically in many parts of Europe and Western Asia, the World Tree has long been considered to be a giant, eternally green ash tree. In Germanic, Norse and Celtic mythology, it appears as the mighty Yggdrasil (known as the Guardian Tree), an ash tree that spans the cosmos. In Norse mythology specifically, Yggdrasil already appears

Giant kapok trees that are found in the tropical forests of Central America represent the Tree of Life for a number of indigenous peoples of the area.

to be present at the beginning of time and is expected to survive Ragnarok, the end of the world, providing the safe haven for a single man and woman who will in time repopulate the planet. The story of Yggdrasil is perhaps the most detailed exposition of any World Tree because it forms part of a series of Norse sagas known collectively as the Prose Edda, written in the thirteenth century, which have remained intact right up to the present day.

In many cultures the ash tree was also seen as the progenitor of mankind itself. In Graeco-Roman mythology, the third, or 'Brazen', race of men are said to be created directly from ash trees by Zeus.[5] In Norse and Celtic origin myths the first man was believed to have

A giant statue of Glooskap, a character from Native American origin mythology who created the world.

A mature ash tree in an English woodland towering over the forest like the mighty Yggdrasil.

been created from an ash log (and a woman from an elm log).[6] Similarly, the Wabanaki (a confederacy that unites five North American Algonquian-language-speaking peoples of the northeast U.S. and southeast Canada) believe that humans were first created from black ash (*F. nigra*) trees.

In one story, a character called Glooskap creates the first people: 'In the beginning *Glooskap* created *Mikumwess*, who were small elves or little men who dwelt among the rocks. *Glooskap* then shot an arrow into the trunk of an ash [black ash] tree and humans came out of the bark.'[7]

There are a limited number of recorded myths linking Native American people with various species of ash trees. In one origin myth for the Athabaskan people in Oregon a character known as *Qawaneca* is said to have created the world.

At first it was dark. There was neither wind nor rain. There were no people or animals. In the middle of the water on a piece of land, sat *Qaweneca*. He sat by his fire of burning cedar. On the edge of the land stood another god. Looking northwards he saw an ash tree. Looking southwards he saw a red cedar. Therefore, ash and red cedar are sacred above all other trees.[8]

The black ash (*F. nigra*) appears to have played a central role in both the daily lives and mythology of a number of Native American peoples.[9]

The idea that the World Tree in Norse mythology is an ash is disputed by some researchers, such as Fred Hageneder.[10] However, what is undisputed is that several species of ash tree are inextricably linked to Indo-European cultures stretching back over thousands of years. Ash trees are referred to specifically in Greek myths such the *Odyssey*, and ash trees also play a central role in the magic and folklore of Celtic cultures.

A representation of Glooskap turning someone into a tree.

A number of North American ash species are central to the mythology of native peoples, such as the Ponca and Omaha.

Norse Cosmology and the Ash Tree

The Norse understanding of the structure of their universe is wonderfully precise, and within their mythology we have the most detailed description of any World Tree, primarily because it is described in a remarkable book containing a collection of sagas known as the *Prose Edda*.

The *Prose Edda* was written in the thirteenth century in Iceland – then considered to be at the very edge of the known world. It is a Norse creation epic that draws heavily on the oral traditions of what is referred to as the 'Viking Age', an era which stretched from approximately AD 780 to 1070.[11] The text (prose) is scattered throughout with references to two distinctive styles of Scandinavian poetry known as *eddic* and *skaldic*. What is remarkable about the *Prose Edda* is that it was written by Snorri Sturluson at a time when most of Northern Europe had already been converted to Christianity and when much of the cultural history was being recorded in Latin rather than in the tongue of the peoples it referred to. Being an outpost far from the central workings of the new dominant religion, Iceland was a land where texts could still be written down in the native language. The *Prose Edda* was faithfully copied over the centuries, and these reprints preserved a rich culture that might have otherwise been lost or, at the very least, reinterpreted to fit in with Christian dogma translation.

The largest section of the *Prose Edda* describes a journey taken by King Gylfi (of Sweden), who hears about the astounding knowledge of the Norse gods, the Æsir, and travels in disguise as an old man called Gangleri to Asgard (home of the Æsir) to learn more. Once there he questions the three manifestations of the god Odin, which are known as High, Just as High and Third. The core of the book is based on the questions Gangleri asks and the responses that these manifestations are able to give him.

The origin of the Norse world was said to have been when the ice from Niflheim in the north spread into the primeval void known as Ginnungagap and combined with fires of Muspellheim in the south. This created the first two beings: a frost giant called Ymir and a cow called Andumla. The cow is then said to have licked the first being into shape from the ice. In turn he had three grandsons – Odin, Vili and Ve – who became gods and who eventually went on to kill the frost giant; from the giant's remains they created the world.

In response to a question from Gangleri in the *Prose Edda* regarding how the world came into existence, High (one of Odin's manfestations) responds:

It is no small matter to be told. They took Ymir [the giant] and moved his body into the middle of Ginnungagap [primeval void or abyss] and made from him the world. From his blood, they made the sea and the lakes. The earth was fashioned from his flesh, and mountains and cliffs from his bones. They made the stones from his teeth and molars and bones that were broken.[12]

The mythology of Norse peoples describes the universe as a tri-centric structure, centred around Yggdrasil and visualized as three large discs, one above the other, each of which is further divided into three, creating nine realms in total.[13] The uppermost realm was called Asgard, which comprised not only Asgard itself but Vanaheimr (home of the gods called Vanir) and Alfheim (home to the 'light elves' and the god Freyr (or Freyer). Midgard was the middle realm where the humans, dwarves and giants resided in Jotunheim, Niddavellir and Svartalfheim respectively. The underworld comprised Hel (Realm of the Dead), Niflheim (World of Fog) and, in some versions of the origin myth, Muspellheim (Land of Fire), where Surt (a giant who ruled over the land of fire and who plays a central role at the end of the world) waits to return at the end of time with his flaming sword. The underworld was presided over by a great serpent or dragon known as Nidhogg.

The trunk and the roots of the celestial ash tree Yggdrasil were seen as connecting and supporting these three distinct levels of existence. Central as Yggdrasil is to Norse myths and cosmology, it arrives with little in the way of explanation other than that it was believed to have existed before the world was created and will survive beyond Ragnarok. It is said that at the end of the world the leaves on Yggdrasil will begin to tremble and one man and one woman will seek sanctuary and protection within its trunk. When all has been destroyed, they will emerge from the tree and repopulate the worlds.

Beyond the uppermost branches of Yggdrasil the sun and moon were said to have been transported across the sky in chariots chased by wolves, and beyond this the star-studded sky was said to have been enclosed by the skull of the giant Ymir (rather like a giant bell jar), which was held up by four dwarves at the points of the compass. It was also said that when the giant Ymir was slain he bled so much that the world flooded and all the ice giants, except one called Bergelmir, were drowned.

Yggdrasil also had three colossal roots that wound in all directions and dimensions. They stretched into Middle Earth (Midgard) and

Heaven (Asgard), as well as down into Niflheim – and each questing root was associated with a particular sacred well. Next to the root that wound its way into Asgard was the Well of Wyrd, around which lived the mysterious Norns. The Norns (similar to the Nemeses in Greek myth) comprised three prophetic women called Urd (Fate), Verdandi (Becoming) and Skuld (Obligation), who governed the fates of all the humans that resided in Midgard. A rainbow bridge (Bifrost) connected the heavens to Midgard, allowing the gods the opportunity to visit the land of mortals on occasion.

Humans, dwarves and giants inhabited Midgard on land enclosed by a wall (said to have been fashioned from the eyebrow or lashes of the giant Ymir), which separated them from the land of the giants and the land of dwarves, and were further encircled by oceans, which in turn were enclosed by a giant serpent biting its own tail. Here another of the roots from Yggdrasil wound its way through Midgard into the land of the giants, close to which is located Mimir's Well. This was guarded by the wise giant Mimir, and a single draft of the precious water from the well was said to grant the drinker wisdom and the gift of prophecy. In the *Prose Edda* Odin is said to have hung himself from an ash tree, also called Yggrasil, beside the well, his side wounded by a spear, until through his delirium the tree transformed into a magnificent eight-legged horse (Slepnir), which enabled him to travel through all nine realms. While hanging from the tree he lost an eye to a raven. In effect, he exchanged one of his 'seeing eyes' for the ability to see into the future via the eye of clairvoyance, thus open-ing himself up to becoming more intuitive. Interestingly, the name Yggdrasil can also be translated as 'Odin's Horse' or 'Gallows Tree'. Ygg was one of the many names under which Odin travelled and *drasil* can be translated as horse. The act of being hanged was sometimes referred to as 'riding the gallows tree', in that the poor unfortunate bobbed up and down as if riding a horse.[14] One of the skaldic poems that are dotted through the *Prose Edda* describes the event: 'Nine whole nights on a wind-rocked tree, wounded with a spear. I was offered to Odin, myself to myself, on that tree that none may ever know.'[15]

The third root plunged deep into the Underworld. Again, the root is associated with a well, Hvergelmir, from which, it was said, all earthly rivers spring. The Underworld was also home to a giant snake or dragon called Nidhogg, who, not content with ripping apart the corpses of the dead and draining their blood, continuously gnawed at the roots of Yggdrasil, alongside a seething tangle of snakes too numerous to count. Hvergelmir was the location of Hel, where people who had led unspectacular lives went when they died. The Underworld, however, was not completely isolated from the other worlds. For example, the squirrel Ratatusk (literally 'drill tooth') was forever oscillating between the highest branches and the deepest roots, carrying news and insults between Nidhogg of the underworld and the giant eagle that sat at the very top of the Yggdrasil, who is said to have created the winds by flapping its vast wings.

The allegoric nature of the World Tree in Norse mythology, in relation to the earthly ash tree, is very pronounced – as was the need to care for and protect it. Even as a cosmogenic tree it was susceptible to potential damage. Snakes constantly gnawed at its roots while, in its branches, deer browsed heavily on the celestial canopy. In Asgard, the great Valhalla (Hall of the Slain) was said to be so tall that the goat known as Heidrun was able to stand on the roof damaging Yggdrasil by biting stalks and leaves off the branches of the great ash tree. Mead was said to have flowed from its udders, which kept 'the slain' in happy intoxication. Even the exposed roots of the World Tree close to the three wells in the three different levels were seen as being in need of care, and they were regularly watered and the exposed bark carefully covered in clay for protection by the three women known as Norns.

Another section of skaldic poem that breaks up the prose of the *Prose Edda* describes the difficulties faced by the great tree:

> The ash *Yggdrasil*
> endures hardship,
> more than men know.

A stag bites from above
and its sides rot;
from below *Nidhogg* gnaws[16]

In the origin myths of a wide variety of cultures around the world it was believed that the very first humans either came from, or were fashioned from, trees and in some cases even inanimate logs. In Norse tradition men were said to be generated from *Ask* (ash) logs and women from *Embla* (elm).

In one of the stories in the *Prose Edda* entitled 'Men Are Created and *Asgard* Is Built. The All-Father Sees Everything' the travelling character Gangleri asks one of the manifestations of Odin (known as High) again 'Where do people come from?' and High replies:

The sons of *Bor* were once walking along the seashore and found two trees. They lifted the logs and from them created people. The first son gave them breath and life; the second intelligence and movement; the third form, speech, hearing and sight. They [*Sons of Bor*] gave them clothes and names. The man was called *Ask* (ash) and the woman *Embla* (elm or vine). From them came mankind and they were given a home behind *Midgard's* wall.[17]

The extended supernatural fable outlined in the *Prose Edda* is familiar to us today as it has heavily influenced numerous famous writers and composers throughout the centuries. For example, Richard Wagner's *Ring Cycle* and J.R.R. Tolkien's *Lord of the Rings* both draw heavily on it. In the character of Odin we can see both of Tolkien's great magicians: Gandalf the Grey and Saruman. It also influenced poets such as Henry Wadsworth Longfellow and W. H. Auden and the Norse saga has been almost subliminally introduced to millions via action heroes of comics and films.

The question is, why was ash chosen to represent the World Tree or Tree of Life? It has been proposed that Yggdrasil or 'Guardian

Tree' may have actually been a yew tree, as opposed to an ash tree, by Fred Hageneder in his book *The Spirit of Trees*. He points out that the literal translation of the *barraskr* tree from the Icelandic sagas was 'evergreen needle ash',[18] which over time has been abbreviated to ash. This, he states, is more in line with a description of a yew tree rather than an ash tree – evergreen and needle-leaved. The yew certainly has many characteristics that could have made it suitable for representing the World Tree of Norse mythology. Yew trees are clothed in dark green needles and are almost miraculous in the way that ancient examples change imperceptibly over the average lifespan of a human. They even fruit in the depths of winter, creating a splash of brilliant red in a landscape often reduced to monochrome in the higher lattitudes of the northern hemisphere. On top of that yew trees provided the best material for the manufacture of longbows and Palaeolithic hunting spears. Similarly, the oak has mythical status in many cultures and is in many ways grander than the ash. It too is considered a possible progenitor of humans, and was linked with the sky gods through its propensity for being struck by lightning more often than most other of the forest trees.[19] Nevertheless, the translation of the 'needle-leaved and evergreen' ash may simply refer to its pointed 'spear-like' leaves. The reference to it being evergreen could simply be artistic licence in the same way that gods are generally considered immortal.

Another issue is that most of the yew tree is highly poisonous to people and animals and apart from its use in longbows its timber is generally less useful than that of other trees such as oak and beech. By contrast, ash provided so much more to the well-being of the Norse and Germanic peoples and was the tree that, in many ways, defined the Bronze Age. For example, it provided the best firewood and was particularly important in the manufacture of significant items such as spears and the handles of other weapons, as well as the tools that helped facilitate settled agriculture. The oldest sagas were said to have originated in the Scandinavian Bronze Age, which lasted for over a thousand years, from about 1600 to 450 BC,[20] a time when the

Swedish and Norwegian farmsteads often have a large ash tree at the centre of the settlement.

yew hunting spears of Palaeolithic peoples had been largely replaced by ash spears. It was also a time when many implements used in the beginnings of settled agriculture were being fashioned from ash timber. The incredible utility of ash particularly in farming, warfare and day-to-day utensils during the Bronze Age would seem to add weight to the fact that the Norse World Tree was likely to be an ash. Further evidence comes from the practice of the Norse hanging people (as sacrifices) from ash trees, mimicking what had happened to Odin when he had sought wisdom from Mimir's Well. In addition, in 'The Ballad of Svipdag' in the *Prose Edda*,[21] it mentions that not only does Yggdrasil provide fodder for animals but its fruits when cooked provide a tonic that helps ensure safe childbirth. All parts of the yew tree (apart from the red flesh surrounding the aril) are deadly poisonous to both humans and animals. In light of the evidence above it is generally accepted in most literature that Yggdrasil was indeed an ash tree.

In Scandinavia today on certain farmsteads there is still a special link between trees and people demonstrated by the tradition of planting a tree at the centre of a farm or village. In Swedish, it is called a *vårdträd*, and in Norwegian a *tuntre*. The tree, often but not always an ash, is considered sacred, and care of it is said to demonstrate respect for the ancestors whose spirits are believed to reside in it.[22] It is said to be a reminder to care for the place where one lives and to generally care for the world around you.

Many villages, recorded in England in the Domesday Book of 1086, were said to have been identified by, and even taken their name from, a single large tree that marked the spiritual centre of the village. For example, villages such as Ashtead, Ashridge, Ashton, Ashburton, Ashmore and Eight Ash Green all took their names from a significant ash tree. Even in the first editions of the Ordnance Survey maps in the early nineteenth century large ash (and other species) trees were recorded as significant landmarks.

Ash Trees and the Celts

The Celts were another European people to whom the ash tree played a significant role in both their daily lives and their mythology. The Celts, as with most historic groups, were not a conveniently homogeneous culture but rather a loose affiliation of peoples that were believed to have originated somewhere in the region of the upper Danube river more than 3,000 years ago, during the Bronze Age.[23]

By the time they had reached their peak, between the fifth and first centuries BC, they had spread across Europe, as far as Scandinavia in the north and Ireland in the west, as well as into parts of Asia Minor. They were prodigious travellers and traders, and their perambulations would have brought them into contact with a number of other cultures of the time with whom they had a shared heritage. By the first millennium Celtic cultures had started to be overrun by the Romans, who were expanding north into Central Europe, and by the Slavic and Germanic peoples, who were expanding into Northern Europe.

A large ash tree at Shane's Castle near Belfast in Northern Ireland.

Rather than disappearing entirely, much of Celtic life became assimilated into the new dominant cultures. (Also, the subjugation of the Celts was not total, with a number of the more remote groups escaping the attentions of the invaders altogether.) As with the Norse culture that continued in Iceland, while most of Europe became Christianized, six Celtic nations managed to retain a degree of independence, preserving much of the culture that had once been almost pan-Northern European. Today the six Celtic nations that continue to be recognized are Brittany (*Breizh*), Cornwall (*Kernow*), Wales (*Cymru*), Scotland (*Alba*), Ireland (*Éire*) and the Isle of Man (*Mannin*).[24] Each has a distinctive language but all share a common heritage. It is through the study of these surviving Celtic cultures and their historic texts that we can get a glimpse of the past and how important the ash tree was to their culture.

The Celts shared much with other European cultures, particularly the Norse and Germanic tribes. The ash tree as the World Tree was a shared concept, as were the trials and sagas of Odin. This is believed to have been because of shared links to early Aryan culture in Asia Minor and also because of regular incursions by the Vikings, particularly in Wales and Ireland. For example, the Welsh Celtic peoples shared the concept of the World Tree, but had their own names for the three levels: Annwn (the lower world), Abred (this world) and Gwynvid (upper world).[25]

While many of the cultural aspects of the ash tree were shared with other European cultures there was a particular emphasis on the magical properties of trees in general, and ash trees in particular, by the Celts. Ash trees were considered particularly magical and one of a trilogy of trees that could constitute a sacred grove – the other two being oak and thorn (hawthorn). In Ireland in pagan times, for example, there were reputed to be five 'Magic Trees', three of which were ash trees. The names of these magical ash trees were the Bile Tortan (Tree of Tortu), Tree of Dathi and the Branching Tree of Usnech, all located in southern Ireland.[26] The Bile Tortan was recorded in later literature as being truly enormous and was said to be some 300 cubits

Two artistic representations of the Norse cosmos comprising three realms supported on a giant ash tree.

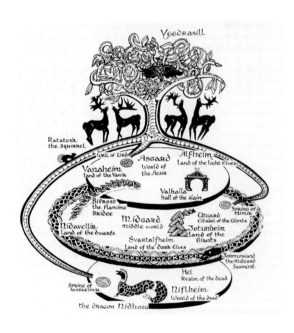

Yggdrasill

Ratatosk
the squirrel

Well of Urd

Asgard
World of
the Aesir

Alfheim
Land of the Light Elves

Vanaheim
Land of the Vanir

Valhalla
hall of the slain

Bifrost
the flaming
Bridge

Spring of
mimir

Midgard
middle world

Utgard
Citadel of the Giants

Nidavellir
Land of the dwarfs

Jotunheim
Land of the
Giants

Svartalfheim
Land of the dark Elves

Jormungand
the midgard
Serpent

Hel
Realm of the dead

Spring of
hvergelmir

Niflheim
World of the dead

the dragon Nidhogg

high and 50 cubits thick. (A cubit is the length of an adult arm or about 18 in./457 mm.) Sadly, these 'magic' trees were said to have been cut down or felled. According to legend, some time around AD 665 the trees were destroyed by Christian missionaries in an attempt to eradicate Druidic practices in Ireland.

Ash trees also played an important cultural role in Celtic Druidic society in healing and superstition. Ash trees were considered to have magical, healing and protective properties. In Celtic or quasi-Celtic folklore ash wood was one of the key materials in the paraphernalia of witches and shamans. Wands, staffs and broomsticks were often crafted from ash wood. The fruits and leaves of ash were also considered to have magical properties. The ash keys (as the bunches of seeds are popularly known), for example, were generally thought to

Charles Knight's illustration of a druid's grove, *c.* 1845.

Ash trees are often struck by lightning, which was one of the reasons why the Celts believed that the wood had special powers and energies.

be an effective protection against the malign sorcery of witches, shamans and evil. The ash was particularly connected with the newborn, as it was in Graeco-Roman folklore. Anna Franklin, a specialist in European folklore, notes that 'ash buds placed in the cradle prevent fairies exchanging a changeling for the child.'[27] In British folklore, the ash is a tree of rebirth – Gilbert White described in his *History of Selborne* published in 1789 how naked children had formerly been passed through cleft pollard ashes before sunrise as a cure for a number of illnesses in a symbolic rebirth. This custom survived in more remote parts of Britain until the 1830s.[28]

The Irish poet W. B. Yeats famously described a 'fairy doctor' who always carried an ash wand with him: 'It is when we come to the fairies and "*fairy doctors*", we feel most the want of some clue – some light, no matter how smoky . . . Why do they fear the hazel tree, or hold an *ash* tree in their hands when they pray?'[29]

The leafing of the ash has also figured in weather lore over the years. 'When the oak comes out before the ash,' a nineteenth-century saying in English Midland counties went, 'there will be fine weather in harvest; but when the ash comes out before the oak, the harvest will be wet.'[30] As recorded by T. F. Thiselton-Dyer in the *The Folk-lore of Plants* (1889),

> Oak before Ash we are in for a splash
> Ash before Oak we are in for a soak.[31]

One of the possible reasons why ash trees were considered special by Celtic peoples, in terms of magic and superstition, may have been derived from that fact that, along with oak, ash trees have a propensity to be struck by lightning. It is probably more a function of their height and the fact that their taproot plunges deep into the water table (unlike most trees) rather than the low internal resistance, as in the case of the oak, that means ash trees are struck so regularly. Robert Graves, author of *The White Goddess*, considered that the ash was a lightning tree simply because Yggdrasil was sacred to Odin. In Norse mythology thunder was said to be the noise of Thor's chariot rumbling across the heavens to confront the giant Hrungnir in battle and that lightning was the fragments of whetstone lodged in his beard. The Celtic god of thunder and fire was Taranis (a northern version of Jupiter), to whom ash, hawthorne and houseleek were all sacred.

However, the observation by the Celts and other cultures of this phenomenon may have given rise to the notion that energy, both good and bad, could be channelled through ash wood, as suggested by the popular, anonymous old English saying, 'Avoid the Ash, / It Draws the Flash.'[32] The ash tree was considered a conduit through which gods such as Taranis could direct their energies.

In many Indo-European languages the words for ash tree and spear are synonymous (*aesc*, *fraxinus* and so on), and ash spears were believed to be a way of directing energy. Similarly, wands and staffs made from ash were also believed to be able to direct energy or power, but of the magical variety. Perhaps the finest example of a decorated Druidic wand was found in Anglesey in the early twentieth century. It was decorated with an intricate silver spiral along its length, and dates back to the first century AD.[33]

Wales has a particularly rich recorded Celtic folklore, where the ash tree is often associated with healing and enchantment. In the Mabinogion – a collection of stories based on oral traditions written down in Wales in the twelfth and thirteenth centuries – one of the most famous characters is a sorcerer known as Gwydion, who uses magic and guile to trick people. He is known to bear an ash staff or wand, a symbol of healing and especially transformation and empowerment in matters of destiny. In Tolkien's *Lord of the Rings* Gandalf's staff is similarly fashioned from ash.

The shafts of witch's brooms were often said to be made from ash because of its association with flight and the ability to travel between realms. In Norse myth, Odin hanged himself from the ash tree next to the well of Mimir for nine days in order to gain knowledge and also obtain the power to travel between realms. In his delirium, the

Druids used ash staffs as part of their magic paraphernalia.

An aerial root in an ancient hornbeam tree looking like the leg of a being about to step out of the hollow.

ash tree eventually transforms into an eight-legged horse that can transport him through all the realms of existence.

The ash finds its way into folklore across the northern hemisphere. For example, in some Germanic traditions, ash trees were considered to be the haunt of powerful witches such as Askafroa. In order to appease her it was said to be necessary to make a donation to her on Ash Wednesday.[34]

Ash was also a tree associated with prophecy. In Norse mythology, the mystical 'Norns', who governed the fate of mortal humans and lived in the upper branches of the great World Tree Yggdrasil, were able to see the future. Similarly, in Celtic cultures it was believed that ash leaves placed under the pillow could facilitate prophetic dreams.[35]

ASH

The Celts shared other folkloric ash traditions and superstitions with multiple cultures. One such notion was that snakes could not endure being near ash trees, or even its cut timber. This was present not only in Norse and Graeco-Roman cultures but among native North American peoples. Similarly, the Celtic cultural observation of connections between the ash tree and the sea was also true of the Greeks and Romans, who believed that the ash tree was linked to the sea god Poseidon.[36] With reference to this the paddles of Celtic boats were generally fashioned from ash, as were the frameworks of their famous coracles. Even during the great exodus of Irish people to the Americas in the nineteenth century, many people boarding the ships destined for the New World were said to be in possession of

Ancient Norse petroglyph with a combination of drawings and runes.

A Swedish runestone standing in a winter landscape.

small crosses made from ash wood as a preventative measure against drowning. The author and poet Robert Graves wrote that 'a descendent of the Sacred Tree of Creevna, an ash, was still standing at Killura in the nineteenth century; its wood was a charm against drowning.'[37]

In the Norse saga Odin was said to have hanged himself from Yggdrasil to gain enlightenment and clairvoyance. At the same time, it was said that when he fell into a trance as he hung on the tree he gained knowledge of Futhark Runes. These were once considered to be the preserve of the gods and comprised an alphabet of runic symbols with which Norse and other Germanic peoples wrote from at least the first century AD. The runes were primarily used to commemorate heroes aand ancestors. As their origin was mythical they were considered able to communicate with the natural and supernatural in the form of protection spells and the like. Similarly, one of the features of Irish Celtic culture was the development of what is known as the 'Ogham' alphabet, which shares many similarities with the runic writing of the Norse.

The Ogham alphabet is a mysterious set of sigils representing fifteen trees and five shrubs that, over time, developed into letters. The Ogham alphabet in Irish mythology was believed to have been created by Ogma, one of the primary gods of Ireland. In one ancient Ogham text it states: 'Ogma, a man well versed in speech and in poetry, invented *Ogham*. The cause of his invention, as proof of his ingenuity, and that this speech should belong to the learned apart, to the exclusion of rustics and herdsmen.'[38]

The Ogham script or writing is quite distinct from that found on Norse runic stones and was believed to have developed independently. It comprises 25 simple sigils with strokes either centred on, or branching off, a central line. This writing is thought to have developed in response to the spread of Latin across Europe, and it originated in Ireland around the first century. It is postulated that it could have been deliberately cryptic to enable Druids and local political leaders to communicate secretively after the Roman invasion of Britain.[39]

Most remaining historic writings using the Ogham alphabet dating from between the first and eighth centuries are generally found carved into stones. Stones inscribed with Ogham script or writing have been found all over Ireland and in parts of Wales, mainland Scotland, England, the Isle of Man and even as far north as Orkney.[40] However, the writing was also believed to have been widely etched onto pieces of wood and bark tablets made from hazel and aspen, known as the 'rods of the *Fili*' where *Fili* means 'poets'. These no longer exist, partly because of the lack of durability of the material on which they were inscribed but also because of the actions of St Patrick, who had nearly two hundred of these historic documents burnt in order to stamp out the influence of the Druids in AD 1400.[41]

According to George Calder (who translated a famous Irish medieval manuscript containing much about Ogham and its history from Gaelic to English, *The Scholar's Primer* (1917)), ash appears as the fifth letter in the Ogham alphabet and was generally represented by a vertical line with five horizontal lines to the right. It was called, variously, *Nion*, *Nuin* and *Nin* (pron. Nee-uhn or Noo-in), and functioned as the letter 'n'.[42]

The ash tree also featured in the Celtic astrological calendar, the Ogham Wheel of the Year, as proposed by the author and poet Robert

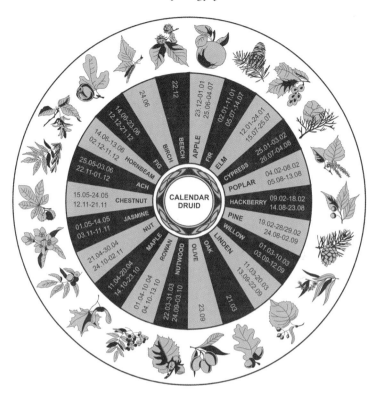

A representation of the tree calendar as observed by Celtic peoples.

Graves in his book *The White Goddess*. The ash represented the third
month of the Ogham calendar and the season of Pisces extending
from 18 February to 17 March – the season of spring floods.[43] Ash
was renowned for its connection to the gods and goddesses of water,
rain and the oceans, particularly Mannanan (equivalent to Poseidon
in Greek mythology).

Some also attribute the tree to the Water Element and others the
Air Element. Others associate the tree with the planetary forces of
the Sun or the Sun in Sagittarius in particular. Ash trees bud in March–
April, which may explain its approximate placement in the Ogham
calendar, according to the work of Graves. Much of our understanding of the Ogham calendar comes from Graves's *The White Goddess* and
he may have invented much of it himself.

The Ash Tree and Graeco-Roman Mythology

The ash tree makes an early debut in Greek mythology and could be said to be almost as central to the Greeks as to the Norse and Germanic peoples. The ash tree is not there at the absolute beginning as in Norse mythology, but it appears in the narrative as early as the Second Order, when the ancient Greek version of the universe is still taking shape.

The First Order is a timeless void where only the binary opposites of darkness and light reign. The Second Order begins when Ouranos (the sky) 'covers' Gaia (the earth),[44] and together they produce a number of offspring and at the very same moment of their union, time begins. They have twelve beautiful children and two sets of hideous triplets – three Cyclopes and three Hecatonchires – creatures with dozens of limbs. Ouranos is disgusted by his mutant offspring, while Gaia loves them all. In the end the behaviour of Ouranos towards Gaia's not-so-beautiful offspring is too much and leads to her plotting revenge.

The sticky white exudate of Mediterranean ash trees, known as manna, leaking from the scored bark.

First, she secretly fashions a sharp sickle and hides it in a cave. Then she persuades one of her own sons, Kronos, to castrate and emasculate Ouranos. This he does. In triumph he picks up what he has cut off and flings his grisly prize, still dripping with blood and semen, across the heavens. From where the blood spills onto the ground new living beings begin to emerge, including the graceful ash tree nymphs called Meliae (*Meliae* is also the word for ash trees) and the not-so-graceful Furies, who dispense rough justice with their ash cudgels.

Despite the overthrow of his father a deep fear grips Kronos, because Ouranos, in his pain, was able to curse him, calling out, 'May your children destroy you as you have destroyed me.'[45] In his paranoia, Kronos exacts a terrible punishment on his own children: every time a new child was born to him and his sister Rhea he immediately swallowed it so that the curse could not be carried out. Eventually, when Rhea gives birth to Zeus, her sixth child, she is so desperate to save him that she wraps a stone in a blanket and tricks Kronos into swallowing this instead. Zeus is spirited off to a cave on the island of Crete, where he is raised by Meliae, who feed him milk from a goat called Amalthea and the sweet waxy exudate called manna that oozes from the trunks of certain types of ash trees.[46]

The idea of the white exudate, manna, as the first food of infants is commonplace in Graeco-Roman writings and subsequently across other European cultures. For example, it was common practice for Scottish highlanders well into the nineteenth century to feed newborn children with a few drops of liquid squeezed from an ash branch. Germanic people were also said to have given ash tree manna to newborns.[47]

A number of species of *Fraxinus* exude a sugary substance that the ancient Greeks called *méli* (honey) or manna. Manna has been harvested from the Bronze Age until today. Although the exudate occurs on various types of ash trees, it is generally associated with *F. ornus*, the 'manna ash', and to a lesser extent the narrow-leafed ash *F. angustifolia*, both of which are to be found in southern Europe, particularly around the Mediterranean.

More than 2,000 years ago the true origins of honey, honeydew and manna were little understood. In ancient Greece honeydew and manna found on ash trees were believed to be the result of a gentle celestial rain from the heavens, which, as such, provided the infant Zeus and the founders of Rome, Romulus and Remus, with their first food. Honey was thought to be the literally heaven-sent manna that had been gathered up and stored by bees.

Pliny the Elder wrote:

Honey comes out of the air, and is chiefly formed at the risings of the stars, and especially when the dog star itself shines forth, and not at all before the risings of the Pleiades, in the periods just before dawn. Consequently at that season at early dawn the leaves of trees are found bedewed with honey and any persons who have been out under the morning sky feel their clothes smeared with damp and their hair stuck together, whether this is the perspiration of the sky, or a sort of saliva of the stars, or the moisture of the air purging itself . . . Falling from so great a height, and acquiring a great deal of dirt as it comes, and becoming stained with the vapour of the earth that it encounters, and moreover having been sipped from foliage and pastures and having been collected in the stomachs of bees – for they throw it up out of their mouths, and in addition being tainted by the juice of flowers, and soaked in the corruptions of the belly and so often transformed, nevertheless it brings with it the great pleasure of its heavenly nature.[48]

The stories of manna or 'honey' trees in Greek mythology are also reflected in other cultures. For example, in Norse mythology it is said that 'honey rained down from the skies' and similar was recorded in Sanskrit writings about *soma*.[49] Whatever the origins of the mythology of the 'honey tree', it seems to have played an important role in Indo-European culture.

The etymology of the word *melía* is interesting. The *Online Liddell-Scott-Jones Greek–English Lexicon* defines the words *melías* as 'manna ash' (*Fraxinus ornus*) and *melí* honey, as well as *méli* as 'sweet gum collected from certain trees'. This links with words and phrases collected by the Greek scholar Hesycheus, who, in the fifth or sixth century AD, compiled what is considered to be the richest lexicon of unusual or obscure Greek words. One of the many references he wrote about ash trees was *melías karpós: tò anthrópon génos* (Seed of ash: race of men).[50] This echoes the mythology of the Norse, who believed that men, *Ask*, were fashioned from ash logs and women from *Embla* (elm). The belief that humans originally came from trees is another almost universal belief across ancient cultures and tribal communities.

The ash tree also features not only in the upbringing of a deity as important as Zeus, but in the creation of early mortal humans, as outlined by the Greek poet Hesiod in his poem *Works and Days*. Hesiod lived sometime between 750 and 650 BC and gives us one of the most complete references to the Ages of Men, described in lines 109–201 of his epic poem.

First came the Golden Age, which was considered to be a perfect era, where the gods freely mingled with the new beings that had been created by Prometheus out of clay. It was said that there was no hunger or disease and that they all lived happily and in harmony. They were said to live to great ages before eventually dying peacefully.

> The gods who own Olympus as dwelling-place,
> deathless, made first of mortals a Golden Race,
> (this was the time when Kronos in heaven dwelt),
> and they lived like gods and no sorrow of heart they felt.
> Nothing for toil or pitiful age they cared,
> but in strength of hand and foot still unimpaired
> they feasted gaily, undarkened by sufferings.
> They died as if falling asleep; and all good things
> were theirs, for the fruitful earth unstintingly bore
> unforced her plenty, and they, amid their store

enjoyed their landed ease which nothing stirred,
loved by the gods and rich in many of herd.[51]

However, despite Zeus' warning, Prometheus gifted to the newly created people the knowledge of fire. Zeus was outraged and first determined that the new beings should be made to suffer. He made them work, to observe the seasons and to live for one hundred years under the dominion of their mothers and then for only a few short years as grown adults. In doing so he created the Silver Age of people. He also vowed to devise a suitable punishment for the Titan Prometheus who had dared to disobey him. Sadly, the Silver Age people also offended Zeus, this time by refusing to worship him and the other gods. As a result, he felt forced to destroy them for their lack of respect for their creators.

Traditional Greek mythology suggests that the Third Age or Bronze Age or brazen people were the first proper race of mortal men and these were said to be created directly from ash trees. The very first of these mortal men is said to have sprung from a cosmogenic ash as the result of the union of an ash nymph (Melia), the daughter of Oceanos, with the river god Inacho, producing a son. He was called Phoroneus. And it was Phoroneus who, in Peloponnesian legend, was not only the first (mortal) man but the first fire-bringer.[52]

In his poem Hesiod relates how Zeus created the third or brazen (bronze) race of men out of ash trees.

But when earth had covered this generation also – they are called blessed spirits of the underworld by men, and, though they are of second order, yet honour attends them also – Zeus the Father made a third generation of mortal men, a brazen race, sprung from ash-trees and it was in no way equal to the silver age, but was terrible and strong. They loved the lamentable works of Ares and deeds of violence; they ate no bread, but were hard of heart like adamant, fearful men. Great was their strength and unconquerable the arms which grew from their shoulders on their strong limbs. Their armour was of bronze, and their houses of bronze, and of bronze were their implements: there was no black iron. These were destroyed by their own hands and passed to the dank house of chill Hades, and left no name: terrible though they were, black Death seized them, and they left the bright light of the sun.[53]

These brazen people were the third in the five ages of men. However, rather than the brazen men being an improvement on the predecessors they were considered flawed by Zeus almost from the very beginning; he watched them constantly fight and quarrel. This age came to an end with the great flood of Deucalion.

Among these brazen men there was quite literally a giant bronze man called Talos, who stood tens of metres tall.[54] He was said to be the son of an ash tree nymph. In some Greek writings Talos was considered to be the last survivor of the brazen race, and was given by Zeus as a gift to Europa to help her protect the island of Crete from invaders. He repelled unwanted visitors by throwing large stones at them, or, if all else failed, heating himself up to staggering temperatures and trapping the victims in a scorching embrace. Despite his size and strength, Talos had one weakness: a single enormous vein that ran from his neck to his ankle. This vein was protected by a thin covering of bronze held in place by a simple bronze nail at his ankle. In one version of the story of Jason and the Argonauts, Medea, a beautiful sorceress, was able to trick Talos into letting her near him, whereupon she pulled the nail out of his vulnerable ankle, thus releasing his life blood, or ichor, killing him instantly.[55]

On the *Argo* with Jason was Peleus, a heroic character in his own right who eventually marries the lovely sea goddess Thetis. Between them they produce another famous character from Greek mythology, Achilles. Achilles had many great adventures and his main battle weapon was a mighty spear. This huge ash spear was said to be handed down to him by his father, who in turn had received it as a wedding gift.[56] The great spear was reputed to have been crafted by the centaur Chiron from the timber of a mighty ash tree that stood on the slopes of Mount Pelion. A metal spearhead was attached, which was polished by Athene, the goddess of battle and wisdom. The ash tree was sacred to Ares,[57] the Greek god of war, presumably because of its usefulness in the manufacture of spears – the main weapons of war in the Bronze Age – and because it was the best material for sword, axe and hammer handles.

four
Useful Ash
❦

Trees have provided a number of key global resources that have enabled people to survive, thrive and develop from prehistoric times to the Industrial Revolution and beyond. They have provided food, medicines and wood from which it was possible to make fire, shelter and a wide range of tools, weapons and other products.

It could be argued that without the ability to manage and harvest useful materials from trees, many of the advances towards civilization as we now know it would have suffered. For example, the development of specialized wooden tools is seen as a necessary precursor for settled agriculture. Equally, many of the main weapons of war between the Bronze Age and the end of the Middle Ages were made either entirely or partly of wood. Even the prized metal objects of ancient cultures – the bronze spear tips, iron arrowheads and sword blades – not only needed wooden handles but were only possible to manufacture with the expert knowledge of woodsmen. These woodsmen would have needed to know how to produce charcoal that burnt at the high temperatures needed to smelt specific metallic ores used in the manufacture of metal objects. In short, without access to resources provided by trees, the peoples of prehistory could never have fashioned more durable substances into tools, which would have greatly impacted settled farming and the development of vital systems of manufacture.

Against this background there are few trees growing in Europe today that have been as important to the development and the

An ancient ash tree in Cumbria, England, which has been pollarded for tree hay.

well-being of mankind over the last 5,000 years as the ash tree. Its timber, leaves, sap and bark have been used by people for thousands of years. Its timber is particularly versatile, providing much of the material crucial to the development of warfare, transport and settled agriculture. This, coupled with the ash tree's almost pan-Indo-European and North American availability, and its ability to colonize and grow to a useful commercial size quickly and to respond well to being managed on a regular rotation, has led it to become one of the most reliable, useful and essential of trees.

Ash Timber

Ash timber has a number of qualities that make it particularly useful: it is strong and flexible and wears smoothly. It is relatively dense, typically around 690 kg (1,520 lb) per cubic metre,[1] which is almost exactly in the middle of the range between balsa (*Ochroma pyramidale*), the lightest timber, and lignum vitae, the heaviest, whose wood is so dense that it will sink in water.

Ash timber also has a number of mechanical properties that help add to its general utility. It is straight-grained and has what is known as a high 'cleavability'. This term refers to how easy it is for the timber to be split and woodworkers over millennia have taken advantage of this property of ash in order to fashion a wide range of items. The reason why timber splits so readily is that as the tree grows, the grain (the alignment of plant cells) tends to be laid down parallel to the direction of tree growth. Unlike wood from trees such as oak (*Quercus sp.*), yew (*Taxus sp.*) and sycamore (*Acer sp.*), which have an interlocking structure to their grain, some species of ash have lines of lateral weakness along the length of the grain. Relative weakness of the timber in a single direction but strength in others means that ash timber is termed 'anisotropic'.

Another mechanical property that adds to the usability of ash timber is what is known as 'toughness'. This is a measure of flexibility, meaning the ability of timber to be put under strain to the point at which it begins to deform internally and then recover the original shape once pressure is released. In this respect, and in terms of withstanding impacts, ash and hickory stand head and shoulders above most other commonly available timbers.

The timber of ash trees is both tough and elastic, due to being 'ring-porous'. This means that within the new timber that is laid down in the trunk each year the tube-like vessels that develop for transporting nutrients and water around the tree are particularly large – often large enough to be seen with the naked eye. However, these large vessels only appear in the early-year growth during the spring and early summer. Timber laid down during the second half of the growing season during late summer and early autumn when there is generally less water available is denser and harder, and has much smaller tubular vessels running through it. This double layer (often visible as bands of different colours) within a single year's growth creates a line of weakness within the structure of the timber, along which the wood can be split. In addition, the porous nature of the wood and the lateral lines dissipate the force of impacts far better than in many other types of wood. This it does by spreading the force of an impact horizontally, meaning that ash timber tends to flake under repeated impacts rather than crack catastrophically.

Ash timber is the woodworkers delight, being able to be worked in the round (whole trunks or branches), split (cleaved) into strips along the straight grain, bent using steam or turned on a lathe. This usability, combined with its unusual combination of mechanical features means that ash timber has been used in the manufacture of a huge array of products – everything from simple spears to composite wooden wheels and from furniture to baseball bats.

While the ash tree has been useful to humans for many thousands of years, it was perhaps at its peak in the seventeenth century. When John Evelyn wrote his remarkable treatise on British trees, *Sylva; or, A Discourse of Forest-trees and the Propagation of Timber*, he recognized its remarkable utility and availability:

The use of Ash is (next to Oak itself) of the most universal. It serves the Soldier . . . and heretofore the Scholar, who made use of the inner Bark to write on, before the invention of paper, etc. The Carpenter, Wheel Wright, Cartwright, for Ploughs, Axel-trees, Wheel-rings, Harrows, Bulls, Oars, the best Blocks and Sheffs as Seamen name them, for drying Herring, no wood like it for tanning Nets . . . excellent for

Detail of the ring pore nature of *F. americana* timber showing the enlarged vessels that give the ash its impact-absorbing quality.

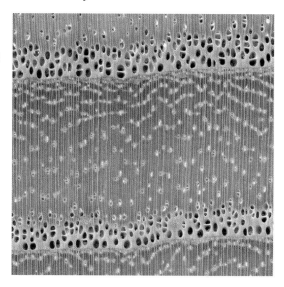

Tenons, and Mortaises; also for the Copper, Turner, Thatcher; Nothing like it for our Pallisade hedges, Hop yards, Pikes, Spars, Handles, Stocks for Tools, Spade Trees etc . . . So as in Peace and War it is the wood in highest request.[2]

William Cobbett (a prominent Member of Parliament and author of the famous book *The Woodland*) in 1825 observed that there were few trees in Britain that had such an extensive range of uses as the ash, and as such it should be better recognized for its economic importance. He goes on to say: 'We could not well have a wagon, cart, a coach or a wheelbarrow, a plough, a harrow, a spade, an axe, or a hammer, if we had no Ash.'[3]

In addition to their incredible utility, many types of ash tree are not only abundant but grow rapidly, reaching commercial size in as little as twenty years. Ash trees also respond well to a variety of management and harvesting techniques, providing a regular and sustainable supply of small-gauge timber, particularly by coppicing and pollarding. These management systems take advantage of the tree's natural response to damage, caused either by natural forces or deliberately by humans. Like a number of other species, the ash tree does not die when cut back but instead tends to respond by putting out vigorous new shoots.

Even as late as the 1930s ash timber was still considered a vital component of the economy in countries such as Britain, leading Mrs Grieves to write in her book *A Modern Herbal*, published in 1931:

> As a timber tree, the Ash is exceedingly valuable, not only on account of the quickness of its growth, but for the toughness and elasticity of its wood, in which quality it surpasses every European tree. The wood is heavy, strong, stiff and hard and takes a high polish; it shrinks only moderately in seasoning and bends well when seasoned. It is the toughest and most elastic of our timbers (for which purpose it was

Ash trees reach commercial size in as little as twenty years.

Traditional ash-wheeled cart. These were used in France until the 1970s
for the lavender harvest.

used in olden days for spears and bows and is still used for
otter-spears) and can be used for more purposes than the
wood of other trees.

It is known that Ash timber is so elastic that a joist of
it will bear more before it breaks than one of any other tree.
It matures more rapidly than Oak and as sapling wood is
valuable. Ash timber always fetches a good price, being next
in value to Oak and surpassing it for some purposes, being
in endless demand in railway and other waggon works for
carriage building. From axe-handles and spade-trees to hop-
poles, ladders and carts, Ash wood is probably in constant
handling on every countryside – for agricultural plenishings
it cannot be excelled. It makes the best of oars and the tough-
est of shafts for carriages. In its younger stages, when it is
called Ground Ash, it is much used, as well as for hop-poles
(for which it is extensively grown), for walking-sticks, hoops,
hurdles and crates, and it matures its wood at so early an age
that an Ash-pole 3 inches [8 cm] in diameter is as valuable
and durable for any purpose to which it can be applied as

the timber of the largest tree. Ash also makes excellent logs for burning, giving out no smoke, and the ashes of the wood afford very good potash.[4]

One of the less desirable features of ash timber is its response to water and damp. In this respect ash cannot compare to oak and chestnut timber in terms of longevity. For example, fence posts made from ash last only a few years before rotting away. This has been well known for thousands of years. However, in terms of looking for evidence of ash wood use in the past, it makes life very difficult for archaeologists and other researchers. It is therefore only in exceptional circumstances that we find ancient ash wood artefacts. Well-known examples include the composite chariot wheels preserved in Tutankhamun's burial chamber in Egypt,[5] and the everyday tableware of ash platters and utensils preserved underwater in anaerobic conditions at the site of Glastonbury lake village in the west of Britain.[6] Much evidence for the use of ash wood in the manufacturing of tools, weapons and so on comes from linguistics, mythology and inference.

In many parts of the northern hemisphere the knowledge of the usefulness of ash trees continues. For example, indigenous peoples in North America continue to employ ash for many different purposes, including firewood, snowshoes, hunting and fishing decoys, lumber for canoe paddles, and baskets.[7] However, the ash tree is not just of historical significance but it is also still commercially important across the northern hemisphere.

Weapons

Ash trees have a long history of providing one of the main materials needed for the development of weaponry. The tendency for ash wood items to rot in damp conditions means that there is less archaeological evidence than for other types of wooden artefacts, making it more difficult to plot the development of weapons made from ash timber over millennia. However, some extraordinary finds have been unearthed and there is strong linguistic evidence to suggest that ash has been an important material for weapons for thousands of years.

The oldest wooden weapons ever unearthed were found in a coal mine in Schöningen, near Hanover, Germany.[8] Radiocarbon dating has reliably aged three spruce (*Picea sp.*) wood javelins to between 380,000 and 400,000 years old. This indicates that pre-human hominids had the skills to produce sophisticated hunting spears, something that was previously only thought to be the prerogative of 'modern man', who appeared less than 150,000 years ago. The next-oldest wooden weapon to have been discovered, dated at 200,000 years old, is a yew (*Taxus baccata*) wood spear, which was found in Clacton, Britain, in 1911.[9] While neither of these examples is made from ash, it shows that the technology to make wooden weapons dates back nearly half a million years and that there were, in all probability, ancient ash weapons which have simply rotted away before being discovered.

One link between ash wood and spears can be traced via the development of Indo-European languages. In a number of proto-Indo-European languages the words for ash and spear are largely interchangeable, such as *os*.[10] The same is true in later languages such as Old Norse and Old Saxon, where *askr* means both ash tree and spear,[11] and Latin, where *ornus* means both too. In Old English and Viking the word *aec* has been used interchangeably to mean both ash tree and spear too. It was also commonly incorporated into terms such as *aesc-plega* meaning 'spear carrier' as in soldier or 'spear play' as in war.[12] A slight variation is the Latin *Fraxinus*, the genus of ash trees, which means 'lance'.

Spears made from ash have certainly been used for long enough for them to have entered the complex mythologies of Indo-European cultures. In Homer's *Iliad* (*c.* 800 BC), for example, the spear wielded by Achilles was handed down from his father Peleus and was said to have been fashioned from timber of the great ash that grew on Mount Pelion by the centaur and oracle Chiron.[13] It was so heavy that only Achilles was strong enough to wield it. In addition to its size

Before the Norman conquest Viking and Celtic settlements were mainly built of ash instead of oak. The interior wooden structure of Viking boats was generally made from ash. Ships at the ancient harbour of Ribe, Viking settlement, Denmark.

and deadly point the spear was also said to have the ability to find its target unerringly. Similarly, Odin, a deity in Norse culture, not only wields a great ash spear but is stuck in the side by one as he hangs himself on the magical ash tree Yggdrasil when seeking ultimate knowledge at Mimir's well. If the mythological evidence of weapons fashioned from ash trees is added to the linguistic evidence, there is a strong likelihood that the history of ash weaponry may stretch back much further than the Bronze Age.

Ash timber has various qualities that would have made it particularly good for fashioning spears and lances: it is tough but not too heavy, it flexes without cracking, and its straight grain makes it easy to work with and particularly resistant to shocks. In addition, ash wood tends not to splinter under impact and the timber actually improves and becomes more comfortable with handling.

Spears are known to have been important in the lives and culture of peoples such as the Vikings. However, they were not simply the main weapon of war but helped denote social status. In the arsenal of the Vikings, which included swords, battleaxes and bows and arrows, it was the spear that was both the most common and the most highly prized weapon. The term *aesc*, which in the Old Norse is pronounced 'ash', can refer to a large, two-handed bladed spear with a 2–3-metre (7–10 ft) ash shaft. In combat, it was used for both thrusting and throwing. Some remarkably well-preserved examples of Viking broad leaf-shaped spearheads have been recovered from the sediment of both the Thames and the Seine where parts of the original ash wood shafts were still present, having been preserved in the anaerobic conditions.[14]

For Vikings, the spear was considered a weapon of status often associated with a person of noble standing such as an ealdorman or an earl, and so valuable that it not only held a special place in their culture and mythology but was often exacted as tribute rather than gold or other precious items.

Alongside the excavated evidence of the importance of ash spears in Viking culture, there is also linguistic and written evidence. For example, in the saga *Beowulf* the word *aesc*, which means both 'ash tree' and 'spear', occurs 25 times: 'Byrhtnoth spoke, and raised his shield, and his wand of slender *ash*, and voiced words: "Do you hear,

Re-enactors of the English Civil War with Royalist pikemen using ash wood pikema at a re-enactment of the Battle of Naseby.

sailor, what these people say? They wish to give you spears as tribute, poisoned point and old sword, such old gear will not help you
in battle.'"[15] There is also the word *aescwiga*, which is used to refer
to soldiers generally but literally translated means 'ash-warrior' or
'spearman'. In a poem of the Battle of Maldon (AD 991), commemo-
rating a famous battle where the Vikings had a notable victory over
the Anglo-Saxons, the noble Bryhtwold 'spoke up, raised his shield
and brandished his *aesc*'.[16] The poem is considered to be the earliest
surviving piece of English literature and the oldest written reference
to the weaponry of the era.

Spears had largely fallen out of favour in European warfare by the
fourteenth century and were largely replaced with pole weapons such
as the halbard and pikema, which both employed an ash wood shaft.
These weapons continued to be used in battle well into the eighteenth century until the advent of the flintlock musket. The military
expert and writer Gervase Markham in his book *Souldiers' Accidence*,
published in England in 1648, extolled the virtues of the pikema. He
wrote that 'every pikema, should have a strong, straight, yet nimble
pike of ash wood, well headed with steel and armed with plates
downward from the head at least four feet, and the full size or length
of every pike should be 15 ft besides the head.'[17]

Spears continued to be used for hunting and fishing across the
world. For example, the Ojibwa of North America used, and continue
to use, white ash (*F. americana*) poles for their fishing spears.

Ash Wood Bows and Arrows

In addition to spears, ash timber was used extensively for the manufacture of ammunition, in the form of arrows, and, to a lesser extent,
bows. Arrows needed to be made from a material that was light
enough to enable the projectile to travel a reasonable distance but
strong and flexible enough to withstand the huge forces upon being
loosed. The physical properties of ash, coupled to the fact that ash
timber was not only readily available but regenerated rapidly and

Bows were manufactured in their millions as the main weapon of war in medieval Europe.

Arrows were often made from ash because of both its availability and its resistance to the great pressures exerted when loosed.

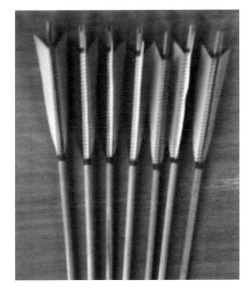

reached a useable size in a relatively short time, made it the ideal material for the manufacture of arrow shafts.

The longbow is a terrifying weapon. The first evidence of longbows dates to 50,000 years ago and several sites in Tunisia,[18] and 23 composite bows were found preserved in Tutankhamun's tomb.[19] Over the centuries the longbow and its lethal projectile continued to develop. Ash was used for the manufacture of bows and arrows in Europe, Asia and North America. For example, it was recorded that green ash (*F. pennsylvanica*) was highly prized for this purpose by native peoples such as the Cherokee, Cheyenne, Dakota, Havasupai, Lakota, Ojibwa, Omaha, Pawnee, Ponca, Potowatomi and Winnebago.[20]

The yew wood longbow reached its peak as the main weapon of European war in the five hundred years from the time of the Norman Conquest of Britain in 1066 until the end of the seventeenth century. It was employed with deadly effect in famous battles such as Crécy (1346) and Agincourt (1415). Here it has been estimated that, at times during these battles, as many as 70,000 arrows were raining down per minute.[21]

The sheer volume of arrows needed for a major battle or war would have been colossal. While there is little data on the actual numbers of ash arrows produced, a good idea of the sheer numbers being produced in readiness for war can be gleaned from English tax records of the fifteenth century. Records show that in February 1417, in any of twenty counties in England, a 'tax' of six feathers from every goose slaughtered had to be sent to the Tower of London in order to provide the fletchers with the material needed for the flights of arrows.[22] This amounts to well over 1 million goose feathers supplied in a single month, which were probably attached to a quarter of a million arrows. Ash was not necessarily the first choice of the military fletcher or any more vital than a number of other woods for the manufacture of arrows. However, a large proportion of the 250,000 ash wood arrows produced in a single month in 1417 would have been made from ash wood. Longbows were the main weapon of war for nearly five hundred years in Europe, which means that billions of arrows in Britain would have been produced – a proportion of which would have been from ash wood.

Of the arrows recovered from the *Mary Rose*, for example, some 80 per cent were made from poplar.[23] However, the sheer availability of ash would have meant that it made a major contribution to the manufacture of what was often termed 'the Devil's finger', as arrows were often called, over hundreds, if not thousands, of years.

Arrows were reported, in ancient Greek and Persian times, as being made of reeds and afterwards of cornel wood, a type of dogwood.[24] However, according to Richard Ascham (1515–1556), a Cambridge lecturer who was also charged with the education of

Queen Elizabeth I, ash was best. Ascham in 1545 wrote the definitive treatise on archery of his time, entitled *Toxophilus or Partitions of Shooting*, which he dedicated to King Henry VIII.[25] In order to reduce the demand on limited supplies of imported yew wood, bowyers in Britain in the time of Henry VIII were directed to make four bows of witch-elm, ash or elm to every one of yew.

What is undisputed is that ash provided one of the finest materials for the handles of weapons such as swords and axes and as such was considered to be the second most important material for making weapons in medieval Europe. It is strong, light and, most importantly, impact-absorbing and does not fracture or splinter in the hand.

In early medieval Europe (*c.* AD 500–1000) weapons not only had a functional role but were often emblematic of a person's status as a warrior, as well as a mark of social rank.[26] Much of what we know about medieval weaponry comes from the excavations of graves and tombs, particularly after AD 500 when the style of funeral rite in Europe changed from largely individual cremations (pyres where the dead were burnt along with their possessions) to burials in specific graveyards. A particularly rich source of artefacts is the burial sites of the Merovingian dynasty in northeastern France, which revealed that the typical weapons of the time comprised a double-edged sword, a single-edged short sword, a shield, an axe, and bows and arrows. An analysis of 316 weapons from 42 burial sites further revealed that the most commonly used wood for weapons at the time was ash, followed by alder and then hazel.[27]

Tools

In many ways, the beginning of the twentieth century marked the end of an extended era where, in Europe and North America, ash had been the pre-eminent wood used to produce tools. A brief dip into a book on British agricultural tools from the turn of the twentieth century reveals an entire lexicon of names for specific instruments that were once widely in use but now have become largely forgotten. For thatched homes in Britain, for example, ash-handled or entirely ash-constructed implements rejoiced in names such as wimble, spud, leggatt, hook and bill, while for peat digging, long ash-handled implements had names such as marking iron, paring iron and slane.[28]

In Britain and other European countries, tools such as rakes varied in design considerably between locations in response to the specific type of meadow they might be used in. Some rakes were made entirely of ash and others were a composite of ash wood handles and hardwood teeth or tines. Forks, spades, scythes, hooks, bills and hammers all generally used ash timber for their handles. Scythes were often custom-made to fit the person using them. Typically, steam-bent ash was used for the elegantly curved handles of scythes, enabling the user to comfortably sweep the blade parallel to the ground. There were also different designs for ladders, hoops, pegs and so on, generally fashioned from local fast-growing ash, between countries and even between counties in Britain.

The list of tools for which ash timber was used can only be described as extensive. Some items were even fashioned from the living ash tree or hedge. For example, items such as ash walking sticks

A traditional British thatching tool called a leggatt.

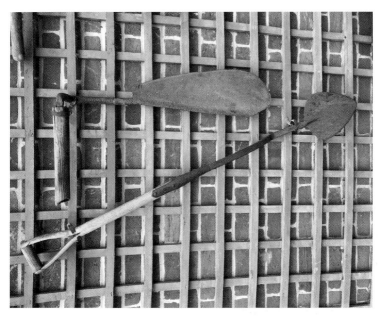

A slane is a specialized sharp spade used to cut peat.

and shepherd's crooks were often partly created while the ash trees were still growing. For walking sticks and umbrellas, for example, the curving handles were often created by woodsmen regularly trimming the apical growing points of coppiced or hedge-grown ash trees, in order to make the stem develop a natural curve.

The knowledge of woodsmen and craftsmen in the past was extensive and sophisticated. Trees were often selected for the manufacture of particular tools based on how and where they were growing. In Britain, ash trees grown on north-facing slopes on carboniferous limestone were considered to provide the best timber for handles. Even poles taken from the same coppice stool base may be considered suitable for different uses. Sadly, much expertise of woodland management and sophisticated knowledge of the finer points of ash timber harvesting have been lost.

Interestingly, in the early twentieth century the demand for ash wood for tool handles and other uses, such as for house building and cabinet making in Europe, led to an over-exploitation of local forest

A selection of scythes and rip hooks all with ash handles.

resources, which stimulated a significant market for the import of American ash (*F. americana*) in particular. This, along with increasing demand within the United States, created concern about American ash timber almost completely running out at a time when it was considered a key national resource for a number of major industries. This was illustrated in a report published on 1 June 1920 by the specialist forester Earle Hart Clapp, who was commissioned by the United States Forest Service. He wrote: 'The demand for ash and hickory handles is so great that manufacturers cannot meet requirements. The export [of ash timber] is even greater than before the war and American handles are being shipped to all parts of the world.'[29] There was a genuine concern that the U.S. would run low on what was considered a strategic material.

The manufacture of ash tools is certainly an ancient skill dating back more than 10,000 years, but most evidence for this tends to be linguistic rather than archaeological in nature because of the susceptibility of ash wood to quickly decompose in the ground. Fortunately, we have a tantalizing glimpse of the sort of tools that were in use some 2,000 years ago from a series of excavations of mining operations in Britain dating from first-century Roman-occupied Britain.

In a report on the excavation of a number of Roman sites, John Robert Travis states that:

> The main types of mining tools used by the Romano-British miner were made of iron and comprised of hammers, wedges, quarry chisels, picks etc. . . . Newstead, Caerleon and Vindolanda . . . The majority of wooden spades so far recovered are of ash.[30]

Ash in Construction and Furnishing the Home

Ash wood has been used for construction in the northern hemisphere since at least Neolithic times (10,000–4500 BC). For example, a series of wooden trackways, known as the Sweet Track, running across the low-lying wetlands of southwest Britain were constructed around 3800 BC to facilitate movement across the marshy terrain. While ash was not the main timber employed, it was nevertheless found to constitute around 10 per cent of the wood used. Interestingly, most of the ash wood used was in the form of small-gauge poles suggesting that most would have been cut from coppiced trees, indicating an early form of tree management.[31] Similarly, ash from coppiced woodland was excavated at the Neolithic Lake village site in Bavaria, Germany, from the middle of the first millennium BC. Here, of the 1,286 timbers recovered, 52 per cent were found to be ash.[32]

Ash was a tree very much associated with the Vikings. So much so that the Anglo-Saxons often referred to the people as *Aescling* – Men of the Ash. Despite ash's lack of durability there is evidence

A number of indigenous peoples of North America employed the bark of
black ash, *F. nigra*, in the construction of their wigwams.

that the Viking's large meeting houses were generally constructed
using massive supporting timbers made of ash. One such example
was excavated from the anaerobic silt on a site close to Dublin dating from the eleventh century.[33] It would appear that in pre-Norman
times, particularly between the tenth and twelfth centuries, the
main wood used in Irish construction was ash. This took the form
of posts and wattle. However, after the twelfth century, oak-framed
Norman-style buildings became the norm, with ash timber relegated
to lesser roles such as panelling, interior joists and basic furniture.
Concurrently it is likely that various types of ash were used by North
American indigenous peoples in the construction of temporary
housing. For example, it is recorded that some Iroquois people used
black ash (*F. nigra*) bark in sheets to cover the frames of their
wigwams.[34]

According to Oliver Rackham, an acknowledged expert on the
history of European woodlands and forests, in medieval Europe
ash was considered very much a secondary or low-grade timber.[35]
However, it was still utilized for internal rafters and joists while the

superstructure of the building would have been more likely made from durable timbers, such as oak. Within the house the small-gauge poles, along with those of a variety of other coppiced trees, were used to form the latticework as part of the wattle and daub walls. It was also used for interior panelling.

In Europe during the Middle Ages ash was a utility material: one which was not only readily available but able to be worked in a wide variety of ways. As such it was likely that ash timber was used for many of the less important household items such as furniture, plates and goblets. Evidence for the use of ash wood in the manufacture of utensils comes from excavations in Cuppers Street in York, which revealed an ash cup factory which operated between AD 950 and 1150.[36] There is also evidence in Britain of ash being used to manufacture utensils going back to Neolithic times. At the base of Glastonbury Tor in the west of England a large quantity of wooden utensils was unearthed during an archaeological excavation – many of which were made from ash wood.[37] It is likely that ash wood was used widely across Northern Europe for many short-lived semi-disposable items. Ash timber is still very much a contemporary material and continues to be used all over the northern hemisphere for the construction and furnishing of buildings.

In Europe, a large industry has built up around the utilization of *F. excelsior* timber, while the same is true of North America, where a number of tall ash tree species such as *F. latifolia*, *F. pennsylvanica*, *F. quadrangulata* and *F. americana* are also widely used in construction. The timber of these is described as having good strength and stiffness and particularly high impact-absorbing properties in relation to other tree timber used in construction, making them ideal for use in flooring, panelling and furniture.

The main ash timber in Europe is provided by the European ash (*F. excelsior*), which has attractive pale (white or pinkish white) wood that is strong, durable, flexible and able to be steam-bent. These features make it particularly suitable for the manufacture of furniture and house interiors. Wood from the European ash is considered easy to work and can be readily turned. The strength and elasticity of ash timber meant that it was used extensively before being largely replaced in the twentieth century by the use of steel for joists and beams.

In North America the timber from the various ash species proved useful to numerous native peoples. The Nipmuc (who lived either side of the Connecticut river) cleverly took advantage of a unique feature of the timber of black ash (*F. nigra*) in order to make woven baskets.[38] Black ash timber lacks the fibres that typically connect growth rings to one another. This enables a technique that involves pounding the soaked timber repeatedly until the weaker spring wood layer is crushed, which allows the darker late-season wood to be peeled off in strips. These strips were then fashioned into baskets traditionally used for berry gathering. Numerous other utensils were made from ash timber by North American peoples, such as ladles

Interior joists and panelling is still made from ash in European and North American homes.

Freshly cut ash logs have a much lower moisture content than most other timber species.

made from white ash by the Iroquois (living in what is now western New York).[39]

Fire

One of the most significant uses of ash trees is for firewood and in many parts of the northern hemisphere it was and still is the primary type of tree used for firewood. This is because ash timber has an almost unique combination of attributes that makes it excellent to burn.

The first definite evidence of fire used by humans apparently comes from the discovery of the remains of hearths and burnt bones near Choukoutien in northern China, which dates from around 400,000 years ago.[40] Despite this long history, it was only around 9,000 years ago that Neolithic peoples in Europe and Asia were able to create fire as and when it was needed. And it was around this time that evidence emerges that ash timber was being regularly used as firewood. In Europe ash starts to appear commonly in fire sites as early as the late Mesolithic in the present-day Netherlands, late Mesolithic–early Neolithic in northwest Belgium and the Neolithic in Britain.[41]

The rise in the use of ash firewood is likely to have been because, as people became expert in making fires, they also began to recognize the particular properties of different types of wood in terms of which would burn best. Ash timber has a number of properties that make it particularly suited to burning, the most crucial of which is its low moisture content when green.

Each type of wood has what is known as its calorific value. This is the amount of energy released per kilogram or tonne of wood expressed as kilowatt hours. The two most important factors determining just how much heat will be produced by a particular wood are its density and its moisture content, and of these it is the moisture content that is most critical. Generally, for a piece of wood to burn well, the moisture content should be around 20 per cent. However, wood when green and freshly harvested contains considerably more than this. The result of burning green timber is that a proportion of the energy released by combustion is used up converting the moisture into steam. In addition, a fire of unseasoned logs can smoke heavily and produce unwanted tars. Therefore, ideally, felled timber would require a certain amount of time to 'season' before becoming firewood.

Ash has an unusually low moisture content in its living timber, around 32 per cent, which means that it will dry more rapidly than most other types of firewood. In addition to its low moisture content ash timber contains relatively high levels of a highly combustible chemical called oleic acid (a fatty acid that occurs naturally in animals and plants), which not only helps the seasoned wood to burn well but enables the wood to burn even when green.

The historical popularity of ash as fuel wood is reflected in an English poem by Celia Congreve from around 1920.[42]

Birch and fir logs burn too fast,
Blaze up bright and do not last.
It is by the Irish said,
Hawthorn bakes the sweetest bread.
Elm wood burns like
churchyard mold,
E'en the flames are cold.
But ash green or ash brown
Is fit for a queen with a
golden crown.

Poplar gives a bitter smoke,
Fills your eyes and makes you choke.
Applewood will scent your room
With an incense-like perfume.
Oaken logs if dry and old,
Keep away the winter's cold.
But ash new or ash old
Is fit for a queen with
a crown of gold.

The particular qualities of ash firewood – being easy and quick to season, relatively smokeless and able to burn while still green – were complimented by the fact that ash trees were not only relatively abundant and quick growing, but they responded well to being managed. In this way sustainable sources of firewood could be developed as far back as the Neolithic by employing techniques such as coppicing and pollarding, which could produce harvestable poles every seven to ten years. Ash trees are also quick to colonize disturbed ground so, as woodlands were cleared and settled agriculture began, the numbers and proportion of ash trees increased where there was human activity or settlements.

Species	Moisture content of green wood
Ash	32%
Sycamore	41%
Birch	43%
Oak	47%
European larch	50%
Douglas fir	51%
Japanese & hybrid larch	51%
Elm	58%
Sitka spruce	61%
Western hemlock	61%
Silver fir	62%
Poplar	64%
Red cedar & lawson cypress	64%
Norway spruce	65%

The relative moisture contents of different types of timber, using data collected by the Forestry Commission, 2010.[43]

Fodder

A feature of trees that is often overlooked in modern agricultural systems is that of providing fodder for animals. Wild grazing animals and domesticated animals such as cattle, sheep, goats and deer, when left to their own devices, will readily include tree bark and foliage as part of their diet. As agriculture developed, and as forests were cleared to make way for fields, trees became more actively managed, either by coppicing or pollarding, in order to provide useful material for both people and animals. Trees were either cut back to the ground (coppicing) or lopped at between 2 and 3.5 m (6½–11½ ft)) above the ground (pollarding) every five to twenty years in order to produce a regular supply of portable timber and foliage to be fed to animals.

Trees such as ash, hazel and willow, when damaged either through natural forces or by human activity, respond with vigorous new growth that is particularly nutritious and appealing to grazers. The techniques of pollarding and coppicing are known to go back at least 3,000 years.[44] Equally it is acknowledged that 'tree hay' has been actively harvested since Neolithic times (for more than 5,000 years). According to the famous nineteenth-century Scottish botanist John Claudius Loudon, ash foliage at the time was used as year-round fodder for sheep and goats in Britain as it was in other countries of Northern Europe such as Norway, Sweden, Denmark and Finland. Ash, along with elm, was one of the most highly prized fodder crops in northern latitudes where hay crops were not possible. Both have high levels of sugar (more than 50 per cent by dry weight) and proteins in their foliage, making them an ideal overwintering food.

Across the temperate regions of Europe most tree and shrub species could be harvested for tree hay. However, if available, the preferred tree species over the last two millennia were ash (*Fraxinus spp.*) and elm (*Ulmus minor*). The reason for this is a combination of elm

Ash trees *F. ornus* grown for both the harvest of manna and also as tree fodder for goats and sheep, Sicily.

and ash foliage being more palatable to animals and having higher levels of minerals and protein than most other tree species. They also have considerably higher levels when compared to a clover-dominated grass meadow. The collection of tree leaves for feeding stock, usually from pollards, is now generally confined to poorer and less inhabited areas of Europe and North Africa, where subsistence farming and traditional herding still take place.

In parts of Europe, such as in the Lake District of northern England, ancient trees known as 'cropping ashes' can still be seen punctuating the landscape. Interestingly the local names for the poles and fodder – *stake and reis* – point to the Germanic and Scandinavian roots of this system and its introduction to the area more than a thousand years ago.

The use of ash (*F. excelsior*) for fodder in northern England was recorded in 1772 by the traveller and naturalist Thomas Pennant, who toured northern England and Scotland in 1769. He noted the use of what were known locally as 'cropping ashes' in the Lake District and how they were vital for the fodder for overwintering animals. In his book *A Tour in Scotland 1769 and a Voyage to the Hebrides* he writes about the competition between hill farmers and the metal smelters in Borrowdale close to the Lake District town of Keswick:

> Observe that the tops of the ash trees were lopped and was informed that it was done to feed cattle in the autumn when the grass was in decline; the cattle peeling off the bark as food. In Queen Elizabeth's time the inhabitants of Colton and Hawkshead fells remonstrated against the number of forges in the county, because they consumed all the loppings and croppings, the sole food of their cattle.[45]

While there are obviously no figures available for the quantities of tree hay harvested by ancient peoples it is possible to get an idea by looking at records of herding communities that harvested tree hay well into the nineteenth and twentieth centuries, such as in Sweden,

Bulgaria, Italy, Morocco and isolated parts of France. In nineteenth-century Sweden, for example, it has been estimated that as many as 9 million trees were pollarded annually to provide tree hay for the country's overwintering livestock.[46]

More recently, research has shown that in the mountains of the High Atlas in Morocco there still exist highly sophisticated multi-purpose management systems of native ash trees, which date back several thousand years, that are for the production of fodder. Here leafy branches of ash are harvested in late summer and fed to small ruminant flocks of sheep and goats when the high altitude and arid, steep slopes of the Atlas Mountains have little in the way of grass for grazing animals. Studies of the Berber peoples in Morocco have shown that a profound ecological knowledge of their local environment has developed over the centuries, in which the ash tree plays a significant part in their survival as pastoralists.[47]

Similarly, in Valdagno in Vicenza, Italy, between the 1920s and 1992, local farms produced around one-third of their annual fodder from managed pollards, mainly from European and narrow-leaved ash (*F. excelsior* and *F. angustifolia*) trees. When comparing the productivity of conventional pasture against ancient wood pasture it was found that on a hectare-for-hectare basis, the wood-pasture system could support one more cow per hectare than a landscape producing only meadow hay.[48]

Transport

There are a number of features of ash timber that have made it suitable for use in modes of transport. It is strong yet reasonably light and very flexible, and can be permanently steamed into a variety of shapes. For this reason, it has been used in the construction of frames for everything from toboggans, horse-drawn coaches and chariots to boats and even aircraft. It was used in the chassis of cars, buses and

Overleaf: 'Cropping ash trees': collecting tree hay or fodder from ash and elm trees has been a Northern European practice for more than 1,000 years.

Composite wheels are made of steamed bent ash sections called fellows.

railway wagons well into the twentieth century. However, perhaps the greatest use of ash timber in transportation dates back more than 4,000 years, to the invention of the composite wheel.[49]

Evidence suggests that the earliest wheels and wheeled vehicles were invented about 5,500 years ago, in or around the Fertile Crescent – an area in the Middle East through which the Tigris and Euphrates rivers drain. The first examples were heavy and cumbersome, generally fashioned from a section of solid trunk. The great breakthrough, which led to the lightweight wheel as we know it today, with hubspokes and a rim, is believed to have occurred in Egypt around 1700 BC. It marked the beginning of a rise in importance for ash timber in transportation.[50]

Around 3,700 years ago Egyptian carpenters developed a composite wheel that was both strong and light, and when used on the chariots, completely changed the nature of warfare of the time. The new-style wheels were made with a hub, spokes and sections of the rim known as felloes or fellows. Complete examples of ash-framed chariots along with their steam-bent wheel rims were discovered in the tomb of the infant king Tutankhamun.[51] What is fascinating is that there was likely to have been little or no ash timber available in Egypt at the time; the timber must have been specifically imported from Syria or beyond. It was the combination of lightness, strength and flexibility that made ash timber the best material for this great step forward in human transport. In addition, the ability of ash timber to be bent by steaming and permanently retain its shape solved the problem of how to retain the curvature of the 'felloes'. There are images on Egyptian friezes clearly showing timber being steam-bent. The new technology of the composite wheel contributed to the expansion of ancient Egypt, due to the terrifying efficiency of the two-person chariot in warfare: having both a driver and an archer allowed the combatants to not only arrive and escape at speed in conflict but to shoot arrows from a relatively safe distance.

Andrew Shortland, in his book *The Social Context of Technological Change: Egypt and the Near East, 1650–1150 BC* (2001), lists the main materials used in an Egyptian chariot contruction:

> the principal materials used for a high-quality chariot were, leather, wood, rawhide, textile, bone, ivory, copper alloy, gold, gypsum plaster faience, glass, stone and glue. The two most important materials were wood and rawhide. For the basic chassis of the vehicle, as well as its axles, wheels, pole and yoke (and sometimes in the case of Tutankhamun, the blinkers) various types of wood were required – particularly ash (*Fraxinus excelsior*), imported from outside Egypt and use for axles, felloes and frame.[52]

From 1500 BC to the Industrial Revolution and beyond, steam-bent ash was used in the manufacture of spoked wheels for wagons, carriages, bicycles and motor cars. In addition, the strength and

Ancient Egyptian relief of a chariot.

flexibility of ash also made it an ideal material for the construction of carriage frames, particularly for carriages conveying people. This was because the flex and give of the frames made for a more comfortable ride for the occupants. The demand for ash by the coach-making companies was phenomenal during the eighteenth, nineteenth and early twentieth centuries – it was estimated that around 40,000 coaches were being made every year in Britain during the 1880s, while in the USA an estimated 1.5 million horse-drawn vehicles were made in a single year in 1900.[53] Ash was used in huge quantities, not only in the manufacture of complex conveyances but in the construction of the simplest of vehicles, such as wheelbarrows and handcarts. The demand for ash reached such a peak in the eighteenth and nineteenth centuries in Europe that large volumes of ash timber were imported from the New World to help meet demand.

Early two-wheeled conveyances, which rejoiced under the names Draisienne, Hobby Horse, Laufmaschine (running machine), Dandy Horse and velocipede (pedestrian accelerator), were generally constructed using an ash wood frame and ash wood wheels with metal attachments.

Even with the advent of the Industrial Revolution, the demand for ash timber in vehicles remained high. Ash wood was originally employed in the manufacture of train carriages and early motor cars because it gave the chassis flex, as well as being simple to manufacture. Metal spoked wheels gradually began to replace wooden ones at the beginning of the twentieth century, but the frames of many of the vehicles continued to be made from ash. Train carriages and the famous London Routemaster buses still had ash frames until the 1930s, and many motor car frames were constructed from ash well into the twentieth century, such as MGs – although these too were gradually replaced by the mass-produced iron and steel chassis. Even as late as the 1960s there were a number of British and American estate cars such as the Morris Traveller that used ash wood frames for the back half of the car. Today, one traditional handmade car manufacturer in Britain – the Morgan car company – still manufactures a traditional, handmade sports car using an ash frame.

Ash has contributed in many other ways over the millennia to the enhanced movement of people. In North America items such as snowshoes and toboggans were fashioned from one of a number of native ash species such as the ubiquitous white ash *F. americana*. Again,

A restored 19th-century hay wagon at the Weald and Downland Living Museum.

the workability of ash contributed to its extensive use. Snowshoes made by the Iroquois people of North America, for example, were made from a single piece of ash split along the grain, and steam-bent into a loop with the two ends bound together by cord – a method that was still in use for the manufacture of tennis racquets up until the 1980s.[54] In the icy expanses of Northern Europe, Asia and North America, sleds and toboggans with steam-bent ash runners were extensively used. Interestingly, in contrast to Scott, the famous British polar explorer, Amundsen, the Norwegian polar explorer,[55] decided to employ more traditional equipment, as used by native peoples of the arctic, in his successful bid to be the first person to reach the South Pole in 1911. This included dog sleds with runners and skis fashioned from ash wood.

North American and European peoples also extensively employed ash in the manufacture of boats as well as for paddles and oars. The

Traditional wooden hiking snowshoes of indigenous North Americans.

Large wooden snow sled from the 1940s.

frames of Welsh and Irish coracles comprised a bent ash frame covered in animal skin, as did many of the traditional boat designs of North American indigenous people such as those made by the Malacite. While the majority of the timber used in the construction of medieval ships in Europe tended to be oak, some parts, such as the internal framework and tenon pegs, were made from ash. In the famous Iron Age Hjortspring boat, excavated off the eastern coast of Denmark and dated to around 350–300 BC, only the lower cross-beams were made of ash, probably because of the lack of durability of ash timber once it becomes wet.[56]

Curiously, in Greek mythology there is a strong connection between the sea god Poseidon and ash. Nearly 3,000 years later the link between the god of the sea and ash was still in evidence, with many of the nearly 2 million people that emigrated from Ireland to the USA between 1820 and 1860 boarding ships carrying a small piece of ash wood about their person as a superstitious precaution against drowning.[57]

One of the lesser known, and maybe more surprising, uses of wood in transportation was in the manufacture of aircraft. While it might be understandable that aircraft from the early twentieth century employed a considerable amount of wood and laminates in their construction, the fact that aircraft used in the Second World War and later also had a considerable number of wooden components may be surprising to many.

In a report by the United States Department of Agriculture and Forest Service in 1941 the uses listed of white ash (*F. americana*) in aircraft are as follows: 'Longerons, propellers, landing gear struts, float ribs, reinforcing for structural members, bent work on wings and fuselages, chines, tail skids, cabane struts, bearing blocks, wing leading edge, floating bulkheads, false keels, control handles and fuselage struts.'[58] This indicates that wood, and ash in particular, was still very

White ash, *F. americana*, is of significant economic value to tool and furniture makers.

Mosquito aircraft made largely from wood, including ash timber.

much considered a necessary strategic material in the manufacture of aircraft well into the 1940s, because without it some aircraft would have been grounded.

Possibly the most famous wooden wartime aircraft was the de Havilland Mosquito fighter-bomber, which is often referred to as the 'Wooden Wonder'. This was a lightweight, highly streamlined twin-engine aeroplane largely made from wooden components. It took less than two years to develop, and between 1940 and 1950 some 7,787 were made.[59] The Mosquito was constructed from several types of wood, primarily English ash (*F. excelsior*), Alaskan spruce and Canadian fir. Remarkably, the Mosquito could take a heavy payload of bombs, was highly manoeuvrable in dogfights and could fly at speeds in excess of 400 mph (640 kph).

Sport

The strength, flexibility and impact-absorbing qualities of ash wood, which once meant that ash was used extensively in the manufacture of weapons of war, is utilized for the paraphernalia of what could be considered necessary for ritualized combat in the form of sport. Ash timber is used in everything from polo mallets, baseball bats, cricket stumps and billiard cues to gymnasium equipment.

In his book *Bat, Ball and Bishop* (1947) Robert W. Henderson suggested that the origins of all games involving a bat and ball can be traced back to the fertility rites of ancient Egyptians.[60] These early games were said to be a re-enactment of the combat between the gods Horus and Set marking the end of winter and the beginning of spring, effectively linking fertility rites and ball games. These Egyptian games and rituals quickly spread across the Mediterranean and became incorporated into a variety of other religions, although the ritual games tended to be suppressed in many areas dominated by Christianity at the time. However, bat and ball games continued to spread into Spain, reaching southern France about 1,000 years ago, at which point they became part of the Christian rituals and ceremonies associated with the spring festival of Easter. Interestingly, when the Catholic clergy in southern France began playing ball games nearly a thousand years ago in the courtyards of the cathedral or in cloisters of monasteries they were effectively observing an ancient Egyptian fertility rite.

Originally, bat and ball games were very much associated with particular religions or belief systems. However, an early version of polo, developed in Persia sometime in the ninth century, spread in such a way as to make it the first truly secular ball game. This is believed to have been closely followed by the French game *la soule*, which gained popularity with peasants as well as religious orders around the turn of the first millennium. Evidence suggests that the games of stoolball, football, hurling, cricket, baseball, golf and billiards all have their origins in *la soule*. The fact that Maya and other people in Meso-America also developed ball games alongside pyramids continues to be a mystery.

Many of the games listed above required wooden bats that could withstand repeated impacts. Over time it became obvious to players of games such as hurling, and later baseball, that ash timber was among the best materials for this purpose. Ash timber, especially when grown quickly, produces one of the best woods for the manufacture of sporting equipment on account of its impact-absorbing, ring-porous nature. Here the enlarged vessels laid down in the early-year section of the annual ring create lines of weakness along which impacts can be dissipated laterally. This means that a bat made from ash will tend to flake under repeated impacts rather than shatter catastrophically, as with some types of wood such as maple. For this reason, ash continues to be the most widely used wood in the construction of hurling sticks, hockey sticks, baseball bats, cricket stumps, snooker cues and gymnasium equipment.

The game that continues to use the largest quantities of ash by far every year is baseball. The wonderfully named Louisville Slugger Company proudly claims to have produced in excess of 100 million baseball bats in its 130-year history, around half of which are estimated to have been crafted from ash timber blanks.[61]

One of the oldest European games for which there are good records is that of hurling in Ireland. The first written record of a game was in relation to the battle of Moytura, which occurred in County Mayo some 1,300 years ago.[62] It was recorded that before the battle a game of hurling was played in which a number of contestants left the field of play badly bruised and in some cases with broken bones.

The object of the game of hurling is to hit a ball (sliotar) between the two goalposts of the opposition using a curved wooden stick (hurley). The hurley resembles a cross between a curved paddle and a hockey stick. The long curved stick developed in order to allow players to run and also in order to have a section parallel to the ground with which to hit the ball or scoop it up.

The ideal stick for hurling, hockey and the like comprises a single shaft made from either a naturally curved piece of ash, or one expertly fashioned by hand. The advantage of a naturally curved stick is that the grain follows the curve, retaining the natural strength of the wood. In this way, the grain is not exposed at the bend, thus reducing a

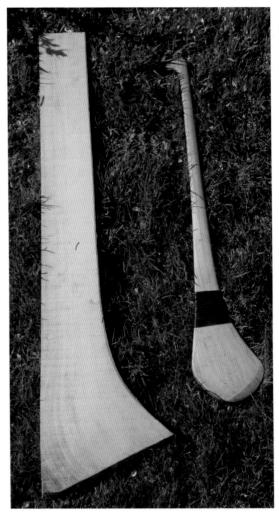

Hurley stick and the
ash blank that it is
crafted from, showing
the flaring of trunk
when the timber is
close to the ground.

potential line of weakness. Players of hurling quickly learnt that a
stick fashioned from a single piece of ash timber cut from close to
the base of the trunk was less likely to break during violent clashes
in the game. The game is popularly known as the 'clash of the ash'.

A report from the Irish Department of Communications, Marine
and Natural Resources outlined the best way to select and prepare
ash timber to make the strongest hurley in order for it to be robust
enough to survive the impacts of this often vicious game:

Ash timber, when grown quickly, is strong, flexible with a good capacity for shock absorbency. For this reason, ash has been traditionally used in Ireland for the production of hurleys. Hurleys are manufactured from the butt log (1.5 m of the bottom of the stem) and from trees at breast height of approximately 30 centimeters. Only fast grown, straight and branch free ash can be used for this purpose. The same shock absorbing qualities make ash suitable for other sports equipment and tool handles.[63]

Sticks used in hockey and the Scottish game of shinty, which is believed to be a permutation of hurling introduced to the Highlands by Irish missionaries, also use the natural curve of ash timber cut from close to the base of the tree.

Like hurling, the stick and ball game most closely associated with it in England, cricket, is based on the ancient French game of *la soule* but with an added element taken from the game of skittles. The British game is likely to be the forebear of baseball.

Linguistic clues help us identify the likely origins of the equipment needed for the game. A wicket, for example, is the name given to a moveable wooden sheep fence often made from a latticework of ash (and other coppice wood) withies, whereas the French word

Professional baseball player in the grand arena at night.

155

cricket simply means club. Perhaps early games involved defending the 'wicket' from a thrown object with a piece of wood. Over time the game developed into not one of simply trying to defend the wicket but trying to hit the projectile as far as possible – beyond the boundary. Eventually, the rough sticks evolved into bats made from specific sections of fast-grown willow. However, the wickets remained resolutely made from ash timber.

Up until the 1980s tennis racquets were also almost exclusively made of ash. Their construction involved the cleaving of a section of ash timber that was between six and twelve annual rings in width, which was then steam-bent into a loop with the straight sections of each end of the hoop glued and bound into position to form the handle, much in the same way as snowshoes are made by native peoples in areas of North America. Being fashioned from a single section of ash wood split along the grain made the racquets extremely durable and able to withstand the regular shock of striking the tennis ball. The last player to use an ash wood tennis racquet in a final at the British tennis tournament Wimbledon was John McEnroe in 1982, when he lost to the sixteen-year-old Boris Becker.[64]

Music

The qualities of ash wood mean that it is suitable for a number of musical instruments. The bottom of a classic violin, for example, is often made of ash. It generally mimicked the sounding board, which was made from fir, in being concave and the same size and shape.

Native Americans had many uses for ash wood. The Sioux, for example, are known to have made carved ash wood whistles and flutes.[65] One of ash's more famous uses in musical instruments is in the construction of electric guitar bodies and to a lesser extent acoustic guitars. Ash guitar bodies are said to have a particular sound quality which is described as a 'bright, cutting tone'. Famous makes of electric guitars such as Fender Stratocasters and Telecasters can be made of Mississippi swamp ash, which can be any one of a number

Many electric guitar bodies are made from a variety of ash timber.

of American ash species that have been growing in waterlogged conditions.[66] Electric guitars with bodies fashioned from alder wood are said to have a darker, richer sound, as in the guitar famously played by Jimi Hendrix. Ash can also be used for making the shells of drums.

Ash: The All-purpose Tree

As has been detailed in this chapter, ash trees have a wide range of properties and features that have made them incredibly useful to many aspects of human development over the last five or more millennia. They provide the best firewood. Their timber has been used to build everything from aeroplanes and chariots to buildings to deadly weapons. It continues to be the best material for making tool handles and for high-impact sports equipment such as baseball bats. It has been important as a fodder tree and has provided medicines in the past and has great potential for the treatment of modern

Green ash, *F. pennsylvanica*, is used widely in the USA for house interiors and furniture.

diseases such as Parkinson's and Alzheimer's in the future. In many ways it was also the tree that helped to facilitate settled agriculture in the northern hemisphere. Because of how quickly it grew and its ability to be managed to produce sustainable sources of timber, the ash tree's population burgeoned alongside that of humanity.

five

The Healing Ash

❦

I n common with many higher plants – relatively complex, mainly flowering plants – those of the genus *Fraxinus* contains within their leaves, bark, roots and seeds an array of complex chemical compounds, many of which have medicinal qualities.[1] As a result the list of disorders for which extracts of ash root, bark, leaf and seed are said to be effective is extensive, including fevers, chronic bacillary dysentery, diarrhoea, meibomian cysts, infantile epilepsy, psoriasis, acute hepatitis, chronic bronchitis and whooping cough.[2] Modern research has shown that the various compounds present in *Fraxinus sp.* can have anti-inflammatory, antimicrobial, antiviral, antioxidant, anti-allergenic, diuretic, wound-healing, and immunomodulatory properties.[3] The medicinal virtues of ash trees have been known for at least 2,000 years and were noted in the writings of the Greek physician Hippocrates (460–377 BC) and also by the Roman scholar Pliny the Elder (AD 23–79). In Europe the German writer, composer and philosopher St Hildegard of Bingen (1098–1179) recorded the use of ash-based remedies for the treatment of rheumatism and gout in the twelfth century.[4]

At the end of the sixteenth century, in the time of the British herbalist and author John Gerard, extracts of ash leaves and bark were already seen as very useful in treating a wide range of ailments. In his famous *Herball* he detailed concoctions of ash tree roots, barks, leaves and seeds which were used to treat everything from fevers (ash bark was used as an alternative to Peruvian bark or quinine), inflammations, warts and liver ailments, and as a purgative and diuretic. He wrote even about their efficacy for snake bites:

> The leaves and bark of the Ash tree are dry and moderately hot . . . the seed is hot and dry in the second degree. The juice of the leaves or the leaves themselves being applied or taken with wine cure the bitings of vipers.[5]

However, at around the same time probably the most comprehensive text on the medicinal properties of ash (*Fraxinus sp.*) ever written was being compiled in Asia. It comes from the era of the Ming dynasty in China with the publication of a most extraordinary medical compendium. Written down by the scholar Li Shizhen and first published in 1578, it has come to be known by a variety of names, including the *Divine Farmer's Materia Medica*, *Chinese Materia Medica* and *The Compendium of Materia Medica*, or *Ben Cao Gang Mu* in Mandarin.

Li Shizhen was a Chinese academic who studied around eight hundred historical medical Chinese texts. This he combined with his experiences of more than thirty years of field work around China in order to compile the most complete phyto-medical review the world had ever seen. The result was a compendium that extends to 53 volumes and forms the basis of much of Chinese medicine today. The core of the book is said to contain 1,892 distinct herbs and 11,096 prescriptions for treating the most common illnesses.[6]

A number of the remedies in the *Materia Medica* and in modern Chinese medicine include the bark of the ash tree, which in Mandarin is known as *qin pi*. Today medicines that are derived from ash bark, which can come from several different species of ash tree, including *F. chinensis* subspecies *rhynchophylla*, are still very important in Chinese medicine because they are effective, readily available and relatively inexpensive.

In the *Compendium* Li Shizhen writes of ash:

F. chinensis is widely used in Chinese and other Asian medical traditions.

It is bitter in flavour and astringent and cold in nature. It goes to meridians of liver, gall bladder, and large intestine. Basic functions are clearing away heat and drying dampness, clearing liver-fire to improve vision, and relieving cough and asthma. Primary indications include damp-heat dysentery, morbid leukorrhea, red painful swollen eyes, eye sores, corneal opacity, lung-heat cough and wheezing.[7]

Interestingly, the knowledge and use of ash trees as medicinal plants seems not to have been restricted to just Asia and Europe but to have existed across the northern hemisphere, as in North America, too, there was a widespread tradition among indigenous communities of using ash bark and leaves for treating common ailments.[8] An idea of the range of medicinal uses to which various *Fraxinus* species have been put can be gleaned from a study of the medicinal plants used by indigenous peoples of Canada published in 1980 by Meredith Black, a Canadian ethnobotanist specializing in aboriginal adaptation in Quebec for the National Museums of Canada.

Black recorded that a tonic made from the bark of ash (*Fraxinus* sp.) known as *ôgmakw* to the Abenaki people was used as an emmenagogue (that is, to stimulate menstrual flow) and also as a cleanser after childbirth. Black also noted that the smoke from burning yellow ash was used to treat earache, and that the Ojibwa people used the inner bark of the black ash or water ash (*F. nigra*), known as *gimak*, to create a tonic which was used to treat sore eyes. The Ojibwa were also recorded as making a tea from a mixture of the inner bark of the green ash (*F. pennsylvanica*), known as *a'gima'k*, mixed with other plants, which was used against weariness and depression.[9]

In other parts of North America, extracts from the leaves and bark of various *Fraxinus* species were used by a diverse range of indigenous people. For example, extracts from white ash were used by the Mesawaki for treatment of snake bites and head vermin such as lice and fleas, while the Iroquois used green ash extract on syphilitic and neck sores, to induce vomiting and for treatment of stomach cramps.[10]

Chinese herbal medicine decoction using bark and other materials.

Ash Trees and Snake Bites

One medicinal use of ash tree leaves and bark which illustrates just how widespread this knowledge was between peoples who were remote from one another is the treatment of snake bites.

The ancient Greek scholars Dioscorides and Hippocrates both noted the efficacy of ash leaves in the treatment of snake bites more than 2,000 years ago. However, probably the most ancient and comprehensive description of ash tree leaves and venomous snakes comes from the writings of Pliny the Elder. In volume VII of his 27-volume *Natural History* he wrote: 'Nothing so sovereign [as ash leaves] can be found against the poison of serpents.' He went on to write:

> Indeed, they [ash trees] are found to be serviceable as an exceptionally effective antidote for snake-bites, if the juice is squeezed out to make a potion and the leaves are applied to the wound as a poultice; and they are so potent that a snake will not come in contact with the shadow of the tree even in the morning or at sunset when it is at its longest, so wide a berth does it give to the tree itself. We can state from actual experiment that if a ring of ash-leaves is put round a fire and a snake, the snake will rather escape into the fire than into the ash-leaves. By a marvellous provision of Nature's kindness, the ash flowers before the snakes come out and does not shed its leaves before they have gone into hibernation.[11]

The assertion by Pliny that ash trees contained an antivenom and that snakes actively avoided ash leaves in particular seems curious at first. However, it appears this belief was common to people across Europe, Asia and even North America. In Cornwall in Britain, it was believed that a blow from an ash stick could instantly kill any serpent.[12] Whereas on the North American continent the Algonquin indigenous peoples of Quebec were recorded by historians placing ash leaves in their boots before walking across areas they believed to be infested with snakes.[13]

In a book on North American medicinal plants by Charles Frederick Millspaugh in the late nineteenth century, a certain Dr Porcher is quoted as saying that:

> The leaves of Fraxinus 'are said to be highly offensive to the rattlesnake that that formidable reptile is never found in land where it grows; and it is the practice of hunters and others having occasion to traverse the woods in the summer months, to stuff their boots or shoes with white ash (*F. americana*) leaves, as a preventative of the bite of the rattlesnake.'

Millspaugh goes on to say, 'My father said that in Orange CO. NY natives defend themselves from snakes by carrying White Ash [*F. americana*] leaves about their person.'[14]

These superstitions were not restricted to North America and Britain, but were probably widespread across Eurasia. And in the United States John Fiske, a 'Lecturee on Philosophy' at Harvard University in the nineteenth century, wrote of the powers of the ash against the serpent:

> The other day I was told, not by an old granny, but by a man fairly educated and endowed with a very unusual amount of good common sense, that a rattlesnake will sooner go through fire than creep over ash leaves or into the shadow of an ash-tree. Exactly the same statement is made by Pliny, who adds that if you draw a circle with an ash rod around the spot on the ground on which a snake is lying, the animal must die of starvation, being effectively imprisoned as Ugolino in the dungeons of Pisa.[15]

More recently biochemists have found that a series of chemicals (secoiridoids and glycosides) isolated from various *Fraxinus* species display what is known as 'anti-complement activation'. This means that treatment of a snake-bite victim with compounds found naturally occurring in ash trees can counteract the venom from snakes such as cobras and pit vipers. It works by preventing 'complement activation' in the normal human serum (the part of the blood excluding clotting agents but including antibodies, antigens, electrolites and so on), thus reducing the debilitating effects of the poison and potentially offering a new snake-bite treatment. Complement activation is a normal part of the body's immuno-defence system that usually reacts to the detection of pathogens with a targeted release of enzymes, often causing inflammation but without damaging the host's own tissues. However, some snake venoms work by over-stimulating the complementary systems, generating large quantities of anaphylatoxins that can be deadly in their own right. Research indicating that chemicals found in *Fraxinus* species could be used to treat snake bites by reducing the chance of anaphalaxis is just one example of the exciting potential medicinal uses to which extracts from ash trees may be put.

It would appear that, while modern biochemists are still distilling, isolating and discovering a range of chemical compounds with potential medicinal value from *Fraxinus* species, peoples across the northern hemisphere, apparently isolated from one another, have had an extensive knowledge of the medical properties of ash trees for hundreds if not thousands of years.

Active Compounds in *Fraxinus*

The main groups of active compounds with known medicinal properties that occur naturally in ash trees include coumarins, secoiridoides and phenylethanoids, polyalcohol – in the form of mannitol – (all of which are a characteristic of *Fraxinus*) and, to a lesser extent, lignans and flavonoids.[16]

Coumarins (including fraxin, fraxetin, scopoletin and to a lesser extent esculin) are what are known as oxygen heterocycles, which can occur in either a free pure form as a colourless crystalline substance or combined with the sugar glucose to form coumarin glycosides. Coumarins are naturally occurring plant chemicals that have a distinctive vanilla-like odour and flavour and as such have been used in the scenting and flavouring of various products including tobacco, alcoholic drinks and fabric conditioners. Coumarins are found in a wide variety of plants that can be distinguished by having a sweet scent, similar to that of new-mown hay, such as meadowsweet. Despite their sweet smell, when in high concentrations coumarins taste bitter and have the effect of suppressing the appetite of grazers and other potential predators.

The most significant contribution to the general modern pharmacopoeia of coumarin is as a precursor in the synthesis of anticoagulant drugs, the most familiar of which being warfarin, which is used to treat patients with a range of blood disorders. Coumarins work by inhibiting the manufacture of the body's vitamin K-dependent coagulation factors in the liver. Unmodified coumarins also have medical value as a treatment for oedema and are known to act as inhibitors to a range of conditions such as hypertension, arrhythmia, osteoporosis and severe pain, and in the prevention of asthma and sepsis and the treatment of HIV and tumours.[17] One of the naturally occurring chemical hydroxycoumarins known as esculin is currently industrially produced from the bark of *F. ornus* in Bulgaria.[18]

Secoiridoides are a subclass of the chemical group iridoids and are found in many species of *Fraxinus*. These compounds have great potential not only in treating snake bites, as mentioned earlier, but in the treatment of a variety of other conditions. For example, four secoiridoids from *Fraxinus chinensis rhynchophylla*, of which two are uniquely characteristic to this species of ash tree (namely insularoside and hydroxyornoside), have been shown to inhibit the activity of pancreatic lipase in digestion. Pancreatic lipase is the enzyme that enables fats to be broken down and absorbed through the gut lining. As an inhibitor of this process it offers great potential in the treatment of obesity.[19]

The bark of *F. angustifolia* is harvested commercially for the pharmaceutical manufacture of drugs in Eastern Europe.

Phenylethanoid glycosides are a relatively newly investigated group of active substances, found in a number of plant genera including *Fraxinus*. These chemicals are of particular interest to the pharmaceutical industry because they have been shown to have antioxidant, immunomodulating, antibacterial, anti-inflammatory, painkilling and neuroprotective properties. Bacterial infections such as *Staphylococcus aureus* can be treated effectively with phenylethanoid glycosides but they have less effect on *E. coli*.[20] However, the general antioxidant properties of these *Fraxinus* extracts means that there are many potential new treatments possible for diseases such as atherosclerosis and Alzheimer's disease.

Lignans are a subgroup of non-flavonoid polyphenols that show great medical potential, as they act as antioxidants as well as defending the body (and tree) against pathogenic fungi and bacteria. Research has shown that lignan derivatives from *F. chinensis rhynchophylla* (particularly sesquilignans) have an inhibitory action on pancreatic lipase and therefore the absorption of fats. This means that they are also seen as having great potential in the treatment of obesity.[21]

Another group of chemicals found in many types of plant, includ‑ ing *Fraxinus*, are flavonoids. Flavonoids are a diverse group of plant chemicals (more than 6,000 types) found in almost all fruits and vegetables. They are responsible (along with carotenoids) for the vivid colours in fruits and vegetables. Flavonoids have increasingly become associated with health benefits. They appear to play a part in longevity, and communities in which there is an unusually high proportion of very old people often have a high level of flavonoids in their diet. It is postulated that they have an ameliorating effect on ageing and may well be associated with the prevention of certain types of cancer and cardiovascular diseases.

Manna and Mannitol

One of the most exciting areas of recent research about the medicinal possibilities of ash trees involves what is known as manna. This is a waxy, white exudate that appears on the branches and stems of ash trees in general, but particularly the 'manna ash' (*F. ornus*), a type of ash tree that predominates in the Mediterranean regions of Europe. As its name suggests it has a connection with the famous biblical phrase 'manna from heaven'. In the Old Testament, it states that manna was '"bread from heaven". It was a "fine, flake-like thing, fine as frost on the ground" and "It was like coriander seed, white, and the taste of it was like wafers made with honey."' (Exodus 16).

The manna of the biblical narrative is likely to have been tamarisk tree (*Tamarisk gallica* var. *mannifera*) manna. This type of manna is exuded from the slender branches of the tamarisk tree in the form of small drops that, first thing in the morning before the temperature rises, are still solid. The Israelites and Roman and Greek scholars thought that manna was heaven-sent, falling as a gentle rain on starry nights. However, the reality is rather less celestial, as the escape of the sap from both tamarisk and ash trees is the result of thousands of tiny puncture marks caused by the insect *Coccus manniparus*, a species of mealy bug found in the Middle East.

A view of forests in Sicily where ash manna has been harvested
as a medicine continuously since the Bronze Age.

Manna, of which the sugar mannitol is the largest component, is
usually harvested from the manna ash (*F. ornus*) and to a lesser extent
the narrowed-leaved ash (*F. angustifolia*). It is an extremely ancient process that dates back at least to the Bronze Age and, in Greek mythology,
at least to the time of Zeus' birth.[22] It was a well-known product to
the ancient Greeks, Romans and Mediterranean peoples generally,
and it has been harvested on the island of Sicily for at least 2,000
years, where today it is still collected around the three small towns
of Castelbuono, Pollina and Cefalù.

Today the plantations of manna ash on Sicily comprise small
trees spaced about 1.5 m (5 ft) apart. Once the trees are eight years
old or more, the bark can be scored with a knife in order to harvest
the manna. When the bark is cut the bitter, often violet-coloured
juice gushes out of the gash. The liquid solidifies almost immediately
upon contact with the air. Harvesting usually takes place between
July and August but the white, sweet, crystalline, exudate can be
collected until as late as the end of October. The chief constituent
of manna is a peculiar, crystallizable chemical compound called man-
nite or manna sugar, which is sweet to the taste and has a peculiar

The scored bark of a manna ash (*F. ornus*) with the manna leaking into a collection vessel.

odour. This accounts for around three-quarters of the volume of the harvested white exudate. Manna also contains a fluorescent chemical called fraxin, which sometimes causes the manna to take on a greenish colour. This sticky white substance is used to make a sugar alternative known as mannitol, which is suitable for diabetics.[23]

Manna has been utilized as an emergency food source and for its medical properties for at least 2,000 years. It is soluble in water and was formerly used as a gentle laxative and now is used mainly as a children's laxative. It has a useful purgative effect, but today is usually prescribed with other purgative compounds, such as senna, rhubarb or magnesia. Its sweet taste means that it is often used as a vehicle for administering other medicines. Its other uses include the treatment of indigestion and constipation; as an antitoxin agent in cases of poisoning; and for coughs, laryngitis, pharyngitis and tonsillitis. Manna sugar is also known as an osmotic diuretic, and has the effect of raising blood plasma osmolality, which in turn increases the flow of water away from body tissues, including fluid in the cranium.[24] Manna sugar can also be used to treat conditions such as increased intracranial and cerebrospinal volume and pressure which can cause conditions like cerebral oedema. Manna sugar also promotes dieresis, helping offset renal failure, and also works as an antiglaucoma treatment.

Perhaps the most exciting research related to the medical effects of manna is in the treatment of brain conditions such as Parkinson's and Alzheimer's. This is because mannitol has been shown to inhibit alpha-synuclein proteins passing through the gut wall. The formation of clumps of alpha-synuclein proteins in the brain (along with the presence of what are known as Lewy bodies) is generally accepted as the main cause of Parkinson's disease.[25] If the protein were unable to traverse the gut wall then it could reduce the associated brain cell death. In addition, it also shows blood–brain barrier inhibitor properties, which also helps reduce the formation of protein clumps in

Ash manna solidifies quickly in hot sunlight.

the brain. Curiously, in research carried out by an Israeli research team, it was found that higher concentrations of mannitol were less effective than much lower concentrations.[26] The potential effectiveness of mannitol in the treatment of brain diseases is considered very promising, especially as mannitol has little in the way of side effects.

F. ornus and *F. angustifolia* both exude a sugary substance known as manna, which has been used medicinally since the ancient Greeks.

Folklore

In the past, the use of ash trees in healing often went beyond the preparation of a simple oral remedy or poultice and into the realms of superstition, where illnesses were often believed to be associated with witchcraft and evil spirits. For example, the openings in trees were often considered doorways between realms, and in Europe, right up until the end of the nineteenth century, it was common practice for people – children in particular – to be 'passed through' holes in trees to cure certain diseases. For example, an article appearing in an 1876 issue of *Report Transactions of the Devonshire Association* stated:

> Passing lately through a wood at Spitchwich, near Ashburton, a remark on some peculiarity in an ash sapling led to the explanation from the game keeper that the tree had been instrumental in the cure of a ruptured infant, and he afterwards pointed out four or five others that had served the same good purpose.[27]

The disease was 'transferred' to the tree by squeezing the person through the opening and 'wiping' the illness off.

Tradition had it that the ill children were passed through the tree three or more times exactly at sunrise. Afterwards the tree was then often bound or plugged with clay and soil. The health of the child from then on was considered to be inextricably linked to the health of the tree in perpetuity, and thus the demise of the tree would be mirrored by the demise of the child.[28]

The ash was also often used as a cure for diseased livestock. For reasons unknown, both livestock and people in Britain were thought to become ill if a shrew were ever to run over them. One of the cures for diseased cattle was that a rod fashioned from ash would be run over their bodies. There were also specific 'shrew ash trees' that were considered effective against diseases too. This is where a poor

A split ash tree was often used for medicinal practice in Europe in the past.

The shrew ash in Richmond Park that was used for the healing
of children until well into the 20th century.

unfortunate shrew would be walled up in the hollow of an ash tree
after which the tree took on magical healing properties.

The remains of one of the most famous and well-documented
shrew ashes stood in Richmond Park in London until as late as 1987.
The tree was reputed to have been ancient and wonderfully gnarled
at the end of the nineteenth century, standing close to the Sheen Gate
on the north side of the park. According to *Magic in the Park*, written

by Marilyn Mason more than 150 years ago, mothers would bring children with whooping cough and other ailments to the tree to take part in a secret dawn ritual led by a 'shrew mother', 'priestess' or 'witch'.[29] This involved the chanting and ministrations of the 'shrew mother' and the passing of the ill child over and under a bar of wood known as the 'witch bar' as the first rays of the morning sun struck the tree.

Another superstition associated with the 'shrew ash' was that a twig or branch from an ash tree with a shrew entombed in its hollow could cure paralysis, effectively reversing the condition that was thought to have been caused by a shrew creeping over the limbs of the afflicted during the night.

Ash also had the reputation of being able to cure warts. There were a variety of methods, some involving transferring warts to the ash tree by wiping the affected areas with bacon, while others required that each wart be pricked by a new pin, which was then stuck into the tree by the afflicted while chanting:

Ashen tree, ashen tree,
Pray buy these warts of me.[30]

Epilogue

❧

Currently there are billions of ash trees spread across the great expanse of the northern hemisphere, quietly going about their business of locking up carbon, ameliorating local climates, cleaning pollution from the air and providing food and shelter for myriad organisms, both large and small. But sadly, like so many other organisms on the planet, the future for ash trees is uncertain.

Over the last 40 to 50 million years, since the first ash trees spontaneously appeared in the southeastern United States, they have expanded right across the northern hemisphere: a slow march from North America into Asia and onward to North Africa and finally into the very western-most fringes of Europe. During that time, they have survived everything a restless Earth could throw at them – volcanic eruptions, glaciations, huge fluctuations in sea levels and great changes in global temperatures. They have survived the thrusting up of jagged new mountain ranges and the erosion to dust of rocks a billion years old. All the time they continued to adapt and evolve into the range of trees that grace our northern hemisphere landscapes today. However, perhaps their biggest evolutionary challenge is facing them right now, as humans, who once heavily relied on and coexisted with ash trees, have helped set in motion a chain of events that could see billions of the world's ash trees disappear in the next century. The very trees that facilitated so much of human development, and which so deeply penetrated the collective

People gather around a five-hundred-year-old ash tree marked as a significant tree on the very first Ordnance Survey maps of the early 19th century in the UK.

psyche of peoples in the northern hemisphere, are now scheduled to be wiped out by our short-sighted actions. In a cruel pincer movement – the combination of emerald ash borer in North America and *Chalara* ash dieback in Europe – ash trees are facing mass extinction.

In Danish folklore, it is said that should ever the ash trees disappear then the world itself would end. A world without ash trees might not be the end of everything, but the northern hemisphere is likely to be a very different place without them. Hope lies in the fact that some genetic lineages display natural resistance to the diseases and that ash trees are fast-growing and exceptional colonizers. Therefore, in the right conditions, they could recover relatively rapidly in geological time but not necessarily in human time. As the great woodland historian Oliver Rackham put it, 'If every ash seed turned into a tree, in two generations the World would not be big enough to hold them.'[1]

Ancient ash trees can support hundreds of species, a number of which
are entirely dependent on ash trees for their survival.

Both F. *excelsior* and F. *angustifolia* are major components of European forests.

A young girl listens to the internal sounds of a wired-up ash tree.

Ash trees are survivors, and there are ways that humans can help. There are biological controls in the form of parasites that can help to reduce the spread of the emerald ash borer. There is also a simple way of preventing *Chalara* ash dieback infecting ash seedlings: by the simple expedient of dipping the exposed roots of ash seedlings into hot water for several minutes the fungus is killed yet the tree survives virtually unscathed. Therefore, we have opportunities to slow down the loss of our ash trees. However, the true solution may require a much more fundamental approach than simply attending to specific infections. Abiotic factors such as climate change, pollution and forest loss and fragmentation are likely to take not just the genus *Fraxinus* but many other species of trees, and indeed types of forest, to the edge of extinction, if the unsustainable way that we treat the planet's finite resources is not addressed soon.

The simple pleasures of sitting under an ash tree or walking through an ash wood on a windy day, with the susurration of tens of thousands of spear-shaped leaves, may be denied to future generations – simple pleasures such as those outlined in a poem by John

Clare, who took it for granted that he could shelter from a sudden rain shower within the bole of one of the local ancient ash trees:

How oft a summer shower has started me
To seek the shelter of a hollow tree
Old huge ash dotterel wasted to a shell
Whose vigorous head still grew and flourished well
Where ten might sit upon the battered floor
And still look round discovering room for more.

Maybe we should look for inspiration – a novel, holistic approach – not only trying to offset the effects of the impending ash tree Armageddon but to learn to appreciate and share the planet with the trees that have facilitated so much of our development. And in

There are still some giant ancient ash trees in Europe, such as the Talley Abbey Ash in Wales which is 7.3 m (24 ft) in girth.

A mature ash tree standing in the dales of North Yorkshire.

looking for new ways of thinking maybe we should take the advice of Robert Graves, who felt that inspiration could be found in the simple yet profound act of listening to the sound of the wind blowing through the leaves in a sacred grove – a sacred grove which in Ireland comprises the trinity of hawthorn, oak and ash.

Glossary

abiotic	Non-living components of ecosystems
anamorphic	Asexual reproductive form in the life cycle of any fungus
androdioecious	Plant that bears only male flowers on some individuals and either bisexual or both male and female flowers on others
anemophilous	Wind-pollinated
angiosperm	Flowering plants whose seeds are encased in fruits. The majority of plants on Earth are angiosperms
anisotropic	Exhibiting properties with different values of strength when measured in different directions
anti-complement activation	Preventing the immune system being activated
ascospores	Single fungal spores formed within an ascus (reproductive cell)
asexual	Reproduction without the fusion of gametes
axillary panicles	A many-branched inflorescence (flower)
biodiversity	The variety of all life forms in either a particular area or habitat on the planet
biotic factors	Living components of an ecosystem
calyx	Sepals of a flower, typically forming a cup-like structure, that encloses the petals and forms a protective layer around a flower in bud
cambium	Thin layer between xylem and phloem that gives rise to secondary growth in stems (trunks) and roots
canker	Fungal infection that damages the bark of trees and is often visible as dark spots
cleavability	The measure of ease with which a piece of timber can be split
clonal	A cell or organism that is genetically identical to that which it is descended from
complement activation	Activation of the immune system

compound leaves	Leaf made up of several or many parts or leaflets all linked to a central stem
coppice stool	Living base from which the new stems arise each time a tree is cut back to the ground (coppiced)
coppicing	Cutting back a tree to ground level on a regular (3–10 year) basis
corona	Crown-like structure of petals
Devil's finger	A reference to the killing power of an arrow
dioecious	Having distinct male and female trees
ditching	Digging drainage ditches
entomophilous	Insect-pollinated
Eocene	Geological period that lasted from approximately 56 million to 34 million years ago
epicormic growth	New shoots that come directly from epicormic buds under the bark of a trunk or branch of a tree
epidermal papillae	Small hairs on the surface (epidermis) of a leaf
epiphytes	A plant that is generally not parasitic but grows on another
fletcher	A maker of arrows
floodplain ecosystems	Ecosystems that are in the floodplain of a river and therefore prone to flooding
frass	Fine powdery refuse and/or excrement of a wood-boring insect
functionally dioecious	A tree that has both male and female reproductive organs but which effectively functions as if it is one sex or another
genus	Main subdivision of a family or subfamily in the classification of organisms
glabrous	Hairy or downy
geo-specific	From particular area of the planet
hermaphrodite	Descriptive of a flower that has both male and female reproductive parts
higher plants	A large group of plants that have developed vascular systems (veins) for the transportation of water and nutrients. Includes flowering plants such as trees and ferns
hurley	The name given to the curved ash wood stick that is used in the Irish game of hurling
imparipinnate	Compound leaf with an odd number of leaflets
inflorescence	The complete flower head of a plant, including bracts, stem and so on
keystone species	Species that either define a particular ecosystem or without which the ecosystem would be impoverished
lanceolate	Shaped like a lance
latewood	Wood that forms late in a tree's growing season and which is visible as a darker part of the annual ring of growth

Miocene	Geological period that lasted approximately 25 million to 10 million years ago
molecular dating	Technique used to age evolutionary events by comparing DNA between lineages
monopodial	Prioritizing growth upwards from a single point
morphology	Study of the structure of plants and animals
necroses	Symptom of a disease where cells die creating dark spots
Neolithic	Final period of the Stone Age lasting approximately from 12,000 to 3,500 years ago
obligate	Having a particular requirement in an ecosystem to survive
obovate	Ovate but with the narrower end downward
Oligocene	A geological period lasting between approximately 34 million and 23 million years ago
orthotropic	The promotion of vertical growth over lateral
ovate	To be oval or egg-like in shape
ovules	Plant structure that gives rise to and contains the female reproductive cells
panicle	A many-branched inflorescence (flower)
petiole	Stalk that joins the leaf to the stem
phloem	Vessel in vascular plants that conducts the sugars and other organic products from the leaves to the stem or trunk
pinnate	Leaves or leaflets arranged on opposite sides of the stem and resembling a feather
pistil	Female organ of a flower
pollarding	The regular cutting back of a tree to just above the browse height (2.5 m) of grazing animals to give a sustainable source of poles and fodder
polygamous	Having both hermaphroditic and unisexual flowers on the same plant or on separate plants of the same species
racemes	A simple inflorescence in which the flowers are borne on short stalks of equal length and at equal spacing
rachis	The stem of a plant
ray parenchyma cells	The living plant tissues that appear as rays in the sapwood of a tree
ring-porous	Timber in which the vessels laid down at the beginning of the growing season are much larger than those of the later part of the year
riparian	The zone that interfaces between a river and dry land
riverplain woodland	Woodland that experiences seasonal flooding
samara	Winged seeds
saproxylic	An organism that feeds on decaying wood
sapwood	The soft recently formed wood that develops between the heartwood and the bark

section	A subdivision of a genus
stamens	The male reproductive part of a plant and which bears pollen
pollarding	The regular cutting back of a tree to just above the browse height (2.5 m) of grazing animals to give a sustainable source of poles and fodder
stomata	A small opening on the surface of a leaf through which gas and water vapour are exchanged
sub-dioecious	Having the majority but not all trees of one sex or another
Sweet Track	A wooden Neolithic trackway, and the oldest man-made footpath in Europe, laid down on the Somerset Levels nearly 6,000 years ago
taxa	Taxonomic categories ranging from kingdom to sub-species
toughness	The mechanical test that determines how strong certain types of wood are relative to each other
turning	Process of crafting a piece of wood into an item using a lathe
understorey	Tier of small shrubs and trees that grow beneath the canopy of a wood or forest
vertex	Upper surface of an insect's head
water table	Level below which the ground is saturated with water
witch's claw	Archaic term for the upturned branches of an ash tree in winter
xylem	Vessel in a vascular plant that conducts water and nutrients upwards from the roots and which forms part of the woody tissues in the trunk of a tree

Timeline

✣

500 million years ago (MYA)	The first land plants appear
360 MYA	The first recognizable trees appear
200 MYA	*Oleaceae* first appear
40–50 MYA	First *Fraxinus* appears in North America
34–23 MYA	Dispersal of *Fraxinus* from America to Asia via land bridges during the Oligocene
11.6–5.3 MYA	Pollen of *F. angustifolia* appears in Europe
400,000 years ago	Oldest wooden spears discovered in Germany found to date to this time
6000–5000 BC	Coppicing and pollarding emerge in the Neolithic age. *Fraxinus* becomes widespread in European forests
3800 BC	Sweet Track in southwest Britain is constructed
1700 BC	Egyptians develop the composite wheel
1600–450 BC	Hunting spears made from yew are replaced by ash spears during the Scandinavian Bronze Age
1332 BC	Chariots with composite wheels made from ash sealed into Tutankhamun's tomb
1272 BC	Battle of Moytur in Ireland, in which ash hurling bats are first recorded

700 BC	Hesiod's *Works and Days* features the ash tree in his description of the creation of early humans
c. 800 BC	In Homer's *Iliad* Achilles' spear is said to have been fashioned from the timber of the great ash of Mount Pelion
AD 50	Dioscorides writing about the medical properties of *Fraxinus*
77	Pliny the Elder reports that ash leaves can be a good antidote to serpents' poison in his *Natural History*
c. 665	The 'magic' trees of Ireland, including Bile Tortan, are destroyed by Christian missionaries attempting to eradicate Druidic practices
800–1100	Viking Era: ash spears are main weapon of war
950–1150	An ash cup factory operates in what is now Cuppers Street, York, England
1000	Decorated ash wand discovered in Anglesey in 1910 dated to this period
1086	The Domesday Book is completed, surveying villages across England and Wales, showing place names deriving from the ash tree, such as Ashtead
1098–1179	St Hildegard of Bingen records the use of ash-based remedies for ailments such as gout
1100s–1200s	The Mabinogion is compiled, this collection of stories including tales of the sorcerer Gwydion and his ash staff
1200s	Snorri Sturluson writes the Norse saga *Prose Edda*
1300	Ash trees first recorded non-fruit trees planted in what is now the UK
1400	St Patrick orders almost two hundred tablets of Ogham writings burnt in order to clamp down on the influence of the Druids

1578	Li Shizhen's *Materia Medica* includes the bark of the ash tree as an ingredient in a number of herbal remedies
Late 1500s	John Gerard records the usefulness of ash for treating a range of maladies in his *Herbal*
1648	Gervase Markham notes in *Souldiers' Accidence* that every pikema should use strong yet nimble ash wood
c. 1650	The largest white ash recorded in North America begins growing
1664	John Evelyn in his discourse on British trees, *Sylva*, recognizes the remarkable nature of the ash tree
1753	Carl Linneaus first describes the genus *Fraxinus*
1772	Thomas Pennant records the use of ash for fodder during his tour of northern England
1789	Publication of Gilbert White's *History of Selborne*
1820–60	Nearly 2 million people emigrate from Ireland to the USA, with many bringing with them small pieces of ash wood as superstitious precautions against drowning on the voyage across the Atlantic
1825	William Cobbett notes the importance of the ash's usefulness to many industries
1880s	Ash baseball bats begin commercial manufacture in USA
Early 20th century	Increasing demand for ash wood in tools and interiors in Europe and the United States
1911	Roald Amundsen, the Norwegian explorer, uses ash wood to fashion sleds and skis for his journey to the South Pole
1930s	London Routemaster buses still have ash frames
1931	Maud Grieve writes in her *Modern Herbal* that the ash is 'exceedingly valuable'.

1940–50	7,787 De Haviland Mosquito planes are made, crafted primarily from English ash
1941	White ash is listed as important by the U.S. Department of Agriculture and Forest Service for use in aircraft
1954–5	J.R.R. Tolkien publishes *The Lord of the Rings* trilogy, with the powerful wizard Gandalf using an ash staff to fight the forces of evil.
1960s	Ash dieback first observed in British hedgerows before spreading across the world. The Morris Traveller uses ash-wood frames in the car design
1979	*F. hubeiensis* first described
1980s	Dutch elm disease causes loss of 30 million elm trees
1982	John McEnroe is the last tennis player to use an ash-wood racquet in the final of the Wimbledon championships
2002	Emerald ash borer is discovered in the USA
2017	BBC *Gardeners' World* viewers select the rose as the most important and influential plant of the last fifty years.

References

Introduction

1 Jesse L. Byock, trans., *The Prose Edda* (London, 2005), p. 18.
2 Darl J. Dumont, 'The Ash Tree in Indo-European Culture', *Mankind Quarterly*, XXXII/4 (Summer 1992), pp. 323–6, www.musaios.com, accessed 16 June 2019.
3 Ibid.
4 Alexander F. Chamberlain, *The Child and Childhood in Folk-thought* (Frankfurt, 2018), p. 132.
5 Melvin Hodge, 'Uses of Plants by the Indians of Missouri River Region', *33rd Annual Report of the Bureau of American Ethnology*, Smithsonian Institution, U.S., Government Printing Office (1919), p. 109.
6 Ibid.
7 Fred Hageneder, *Yew: A History* (Stroud, 2007), p. 104.
8 Richard A. Gabriel, *Thutmose III: A Military Biography of Egypt's Greatest Warrior King* (Lincoln, NE, 2009), p. 13.
9 'European Wood Pasture (Silvopasture)', Case Studies (2012), www.coppiceagroforestry.com, accessed 18 July 2019.
10 Thomas A. Kinney, *The Carriage Trade: Making Horse-drawn Vehicles in America* (Baltimore, MD, 2004), p. 34.
11 K. N. Venugopala, V. Rashmi and B. Odhav, 'Review of Natural Coumarin Lead Compounds for Their Pharmacological Activity', *BioMed Research International* (March 2013), Article id. 963248.
12 'Sweet Approach', www.clinicrowd.info, accessed 9 April 2019.
13 Peter A. Thomas, 'Ashes to Ashes', www.keele.ac.uk, 23 March 2016.

1 The Botany of *Fraxinus*

1 Carl Linnaeus, *Species plantarum* (Stockholm, 1753).
2 Eva Wallander, 'Systemics of *Fraxinus* (Oleaceae) and the Evolution of Dioecy', *Plant Systematics and Evolution*, CCLXXIII (2008), pp. 25–49.

3 A. Lingelsheim, 'Oleaceae-Oleoideae-Fraxineae', in A. Engler, *Das Pflanzenreich*, IV/243 (Leipzig, 1920), pp. 1–65.

4 Robert E. Farmer, *Seed Ecophysiology of Temperate and Boreal Zone Forest Trees* (Boca Raton, FL, 1997), p. 47.

5 Wallander, 'Systemics of *Fraxinus*'.

6 Ibid.

7 Ibid.

8 Ibid.

9 Ibid.

10 W. Carter Johnson, 'Estimating Dispersibility of *Acer*, *Fraxinus* and *Tilia* in Fragmented Landscapes from Patterns of Seedling Establishment', *Landscape Ecology*, 1/3 (1988), pp. 175–87.

11 Ibid.

12 Ivanka N. Kostova, 'Chemical Components of *Fraxinus Ornus* Bark: Structure and Biological Activity', *Studies in Natural Products Chemistry*, XXVI (2002), pp. 313–49.

13 'Plants for a Future, *Fraxinus quadrangulata*', https://pfaf.org, accessed 19 April 2019.

14 Monumental Trees, www.monumentaltrees.com, accessed 19 April 2019.

15 Ibid.

16 Ibid.

17 Ibid.

18 'Field Guide to the Ash Trees of North-eastern United States', www.nybg.org, accessed 18 June 2019.

19 P. Wardle, 'Biological Flora of the British Isles: *Fraxinus excelsior* L.', *Journal of Ecology*, XLIX (1961), pp. 739–51.

20 Ibid.

21 J. P. Grime and R. Hunt, 'Relative Growth-rate: Its Range and Adaptive Significance in a Local Flora', *Journal of Ecology*, LXIII (1974), pp. 393–422.

22 N. M. Collins, ed., *The Conservation of Insects and Their Habitats* (San Diego, CA, 1989), p. 159.

23 Wallander, 'Systemics of *Fraxinus*'.

24 R. H. Ree and S. A. Smith, 'Maximum Likelihood Inference of Geographic Range Evolution by Dispersal, Local Extinction, and Cladogenesis', *Systematic Biology*, LVII/1 (2008), pp. 4–14.

25 V. B. Call and D. L. Dilcher, 'Investigations of Angiosperms from the Eocene of South-eastern North America: Samaras of *Fraxinus wilcoxiana* Berry', *Review of Palaeobotany and Palynology*, LXXIV/3–4 (1992), pp. 249–66.

26 Damien Daniel Hinsinger et al., 'The Phylogeny and Biogeographic History of Ashes (*Fraxinus*, Oleaceae) Highlight the Roles of Migration and Vicariance in the Diversification of Temperate Trees', *PLOS ONE*, VIII/2 (2013).

27 Call and Dilcher, 'Investigations of Angiosperms from the Eocene of South-eastern North America'.

28 Ree and Smith, 'Maximum Likelihood Inference of Geographic Range Evolution by Dispersal, Local Extinction, and Cladogenesis'.

29 Hinsinger et al., 'The Phylogeny and Biogeographic History of Ashes (*Fraxinus*, Oleaceae)'.

30 Wallander, 'Systemics of *Fraxinus*'.

31 Eva Wallander, 'Systematics and Floral Evolution in *Fraxinus* (Oleaceae)', *Belgische Dendrologie Belge* (2013), pp. 38–58.

32 Wallander, 'Systemics of *Fraxinus*'.

33 Ibid.

34 'U.S. Forest service', www.fs.fed.us, accessed 19 April 2019.

35 Wallander, 'Systemics of *Fraxinus*'.

36 George A. Petrides, *Trees of the American Southwest* (Mechanicsburg, PA, 2005), p. 30.

37 '*Fraxinus nigra*, Black Ash', www.borealforest.org, accessed 19 April 2019.

38 Wallander, 'Systemics of *Fraxinus*'.

39 '*Fraxinus chinensis* Roxb', treesandshrubsonline.org, accessed 19 April 2019.

40 Wallander, 'Systemics of *Fraxinus*'.

41 '*Fraxinus chinensis* Roxb'.

42 '*Fraxinus hubeiensis*', www.revolvy.com, accessed 19 April 2019.

43 Oliver Rackham, *The Ash Tree* (Toller Fratrum, 2014), p. 24.

44 J. B. Faliński, 'Vegetation Dynamics in Temperate Lowland Primeval Forest, Ecological Studies in Białowieża Forest', *Geobotany*, VIII (1986), pp. 1–537.

45 Wallander, 'Systemics of *Fraxinus*'.

46 D. Dobrowolska et al., 'A Review of Ash (*Fraxinus excelsior* L.): Implications for Silviculture', *Forestry*, LXXXIV/2 (2011), pp. 133–48.

47 'Distribution Map of Common Ash', www.euforgen.org, accessed 23 September 2019.

48 Lindsay Maskell et al., 'Distribution of Ash Trees (*Fraxinus excelsior*) in Countryside Data', Centre for Ecology and Hydrology, www.countrysidesurvey.org.uk, accessed 19 April 2019.

49 Ibid.

50 Monumental Trees, www.monumentaltrees.com, accessed 19 April 2019.

51 Dobrowolska et al., 'A Review of Ash (*Fraxinus excelsior* L.)'.

52 Alfas Pliûra and Myriam Heuertz, 'Technical Guidelines for Genetic Conservation and Use for Common Ash (*Fraxinus excelsior*)', *European Forest Genetic Resources Programme*, www.euforgen.org (2003), accessed 19 April 2019.

53 Yann Vitasse et al., 'Leaf Phenology Sensitivity to Temperature in European Trees: Do Within-species Populations Exhibit Similar Responses?', *Agricultural and Forest Meteorology*, CXLIX/5 (2009), pp. 735–44.

54 Eva Wallander and FRAXIGEN, *Ash Species in Europe: Biological Characteristics and Practical Guidelines for Sustainable Use* (Oxford, 2005), p. 51.

55 Eva Wallander, *Evolution of Wind-pollination in Fraxinus (Oleaceae) – An Ecophylogenetic Approach*, PhD thesis, Göteborg University, Sweden (2001).

56 Ophir Tal, 'Flowering Phenological Pattern in Crowns of Four Temperate Deciduous Tree Species and Its Reproductive Implications', *Plant Biology*, XIII/1 (2011), pp. 62–70.

57 Pierre Binggeli and James Power, 'Gender Variation in Ash (*Fraxinus excelsior* L.)', www.mikepalmer.co.uk, accessed 19 April 2019.

58 Ophir Tal, *Comparative Flowering Ecology of Fraxinus excelsior, Acer platanoides, Acer pseudoplatanus and Tilia cordata in the Canopy of Leipzig's Floodplain Forest*, PhD thesis, University of Leipzig, Germany (2006).

59 Ibid.

60 Ibid.

61 J. P. Grime, J. G. Hodgeson and R. Hunt, *Comparative Plant Ecology: A Functional Approach to Common British Species* (Dordrecht, 2007).

62 G. Kerr, 'Silviculture of Ash in Southern England', *Forestry*, LXVIII/1 (1995), pp. 63–70.

63 Wardle, 'Biological Flora of the British Isles: *Fraxinus excelsior* L.'.

64 Per-Göran Tapper, 'Irregular Fruiting in *Fraxinus excelsior*', *Journal of Vegetation Science*, III/1 (2009), pp. 41–6 ·

65 P. Collin and P. M. Badot, 'Le point des connaissances relatives à la croissance et au développement du Frêne commun (*Fraxinus excelsior* L.)', *Acta Botanica Gallica*, CXLIV/2 (1997), pp. 253–67.

66 M. C. Dacasa Rudinger and A. Dounavi, 'Underwater Potential Germination of Common Ash Seed (*Fraxinus excelsior* L.): Originating from Flooded and Non-flooded sites', *Plant Biology*, X/3 (2008), pp. 382–7.

67 M. Mund et al., 'The Influence of Climate and Fructification on the Inter-annual Variability of Stem Growth and Net Primary Productivity in an Old-growth, Mixed Beech Forest', *Tree Physiology*, XXX/6 (2010), pp. 689–704.

68 Ibid.

69 H. Pfanz et al, 'Ecology and Ecophysiology of Tree Stems: Corticular and Wood Photosynthesis', *Naturwissenschaften*, LXXXIX/4 (2002), pp. 147–62.

70 Ibid.

71 Ibid.

72 H. W. Buston and H. S. Hopf, 'Note on Certain Carbohydrate Constituents of the Bark of Ash (*Fraxinus excelsior*)', *Biochemical Journal*, XXXII/2 (1938), pp. 44–6.

73 H. Cochard et al., 'Developmental Control of Xylem Hydraulic Resistances and Vulnerability to Embolism in *Fraxinus excelsior* L.: Impacts on Water Relations', *Journal of Experimental Botany*, XLVIII/3 (1997), pp. 655–63.

74 Ibid.

75 Ibid.

76 S. H. Clarke, 'Recent Work on the Relation between Anatomical Structure and Mechanical Strength in English Ash', *Forestry*, IX (1935), pp. 132–8.

77 Wardle, 'Biological Flora of the British Isles: *Fraxinus excelsior* L.'.

78 Ibid.

79 Ibid.

80 Ibid.

81 B. Moe and A. Botnen, 'A Quantitative Study of the Epiphytic
    Vegetation on Pollarded Trunks of *Fraxinus excelsior* at Havrå, Osterøy,
    Western Norway', *Plant Ecology*, CXXIX/2 (1997), pp. 157–77.
82 L. Davies et al., 'Diversity and Sensitivity of Epiphytes to Oxides
    of Nitrogen in London', *Environmental Pollution*, CXLVI/2 (2007),
    pp. 299–310.
83 R. J. Mitchell et al., 'The Potential Ecological Impact of Ash Dieback in
    the UK', *Joint Nature Conservation Committee Report* (2014), p. 483.
84 Ibid.
85 Ibid.
86 M. T. Jönsson and G. Thor, 'Estimating Coextinction Risks from
    Epidemic Tree Death: Affiliate Lichen Communities among Diseased
    Host Tree Populations of *Fraxinus excelsior*', *PLOS ONE*, https://doi.org
    (2012), accessed 19 April 2019.

2 The Threatened Ash

 1 Peter A. Thomas, 'Ashes to Ashes', www.keele.ac.uk, accessed 18 July 2019.
 2 L. V. McKinney et al., 'Presence of Natural Genetic Resistance in *Fraxinus
    excelsior* (Oleaceae) to *Chalara fraxinea* (Ascomycota): An Emerging
    Infectious Disease', *Heredity*, CVI/5 (2011), pp. 788–97.
 3 A. Gross, T. Hosoya and V. Queloz, 'Population Structure of the Invasive
    Forest Pathogen *Hymenoscyphus pseudoalbidus*', *Molecular Ecology*, XXIII/12
    (2014), pp. 2943–60.
 4 S. K. Hull and J. N. Gibbs, 'Ash Dieback: A Survey of Non-woodland
    Trees', *Forestry Commission Bulletin* (1991), pp. 1–32.
 5 T. Kowalski and O. Holdenrieder, 'Pathogenicity of *Chalara fraxinea*', *Forest
    Pathology*, XXXIX/1 (2009), pp. 1–7.
 6 T. Kowalski, '*Chalara fraxinea* sp. nov. Associated with Dieback of Ash
    (*Fraxinus excelsior*) in Poland', *Forest Pathology*, XXXVI/4 (2006), pp. 264–70.
 7 V. Queloz et al., 'Cryptic Speciation in *Hymenoscyphus pseudoalbidus*', *Forest
    Pathology*, XLI/2 (2011), pp. 133–42.
 8 A. Gross et al., '*Hymenoscyphus pseudoalbidus*, the Causal Agent of European
    Ash Dieback', *Molecular Plant Pathology*, XV/1 (2014), pp. 5–21.
 9 J. Schumacher, R. Kehr and S. Leonhard, 'Mycological and Histological
    Investigations of *Fraxinus excelsior* Nursery Saplings Naturally Infected
    by *Chalara fraxinea*', *Forest Pathology*, XL/5 (2010), pp. 419–29.
10 Kowalski, '*Chalara fraxinea* sp. nov. Associated with Dieback of Ash
    (*Fraxinus excelsior*) in Poland'.
11 R. J. Mitchell et al., 'The Potential Ecological Impact of Ash Dieback in
    the UK', *Joint Nature Conservation Committee Report* (2014), p. 2.
12 J. Stokes and G. Jones, 'Ash Dieback: An Action Plan Toolkit', Tree
    Council, www.treecouncil.org.uk, accessed 9 August 2019.
13 See Tove Hultberg et al., 'Ash Dieback Risks an Extinction Cascade',
    Biological Conservation, CCXLIV (2020), available at www.sciencedirect.com,
    accessed 14 November 2020.

14 I. Machar, 'Floodplain Forests of Litovelské Pomoraví and Their
    Management', *Journal of Forest Science*, LIV (2008), pp. 355–69.
15 M. Pautasso et al., 'European Ash (*Fraxinus excelsior*) Dieback:
    A Conservation Biology Challenge', *Biological Conservation*, CLVIII (2013),
    pp. 37–49.
16 Mitchell et al., 'The Potential Ecological Impact of Ash Dieback',
    pp. 157–8.
17 Oldrich Cizek and Martin Konvicka, 'What Is a Patch in a Dynamic
    Metapopulation? Mobility of an Endangered Woodland Butterfly,
    Euphydryas maturna', *Ecography*, XXVIII/6 (2005), pp. 791–800.
18 'P&D Management and New Threats, Kuppen's View', www.trees.org.uk,
    accessed 24 July 2019.
19 Kent F. Kovacs et al., 'Cost of Potential Emerald Ash Borer Damage in
    U.S. Communities, 2009–2019', *Ecological Economics*, LXIX/3 (2010),
    pp. 569–78.
20 Ibid.
21 'Invasive Pest Risk Maps: Emerald Ash Borer (EAB) – *Agrilus planipennis*
    (Fairmaire)', https://data.nal.usda.gov, accessed 30 July 2019.
22 Kovacs et al., 'Cost of Potential Emerald Ash Borer Damage'.
23 Ibid.
24 Karen Cladas, 'Effectiveness of Emerald Ash Borer (*Agrilus planipennis*)
    Trap Placement in Relation to Forest Edges', Masters thesis, Michigan
    Technological University (2016).
25 Ibid.
26 Ibid.
27 D. T. Williams et al., 'Distribution, Impact and Rate of Spread of
    Emerald Ash Borer *Agrilus planipennis* (Coleoptera: Buprestidae) in the
    Moscow Region of Russia', *Forestry*, LXXXVI/5 (2013), pp. 515–22.
28 S. S Izhevskii and E. G. Mozolevskaya, '*Agrilus planipennis* Fairmaire in
    Moscow ash trees', *Russ J Biol Invasions*, I (2010), pp. 153–5.
29 H. M Griffiths, W. A Sinclair, C. D Smart and R. E Davis, 'The
    Phytoplasma Associated with Ash Yellows and Lilac Witches' Broom:
    "Candidatus Phytoplasma fraxini"', *International Journal of Systematic
    Bacteriology*, XLIX (1999), pp. 1605–14.
30 J. R. Clark, 'Adaptation of Ash (*Fraxinus Exelsior* L.) to Climate Change',
    www.scottishforestrytrust.org.uk, accessed 21 March 2020.
31 M. Broadmeadow and S. Jackson, 'Growth Responses of *Quercus petraea*,
    Fraxinus excelsior and *Pinus sylvestris* to Elevated Carbon Dioxide, Ozone and
    Water Supply', *New Phytologist*, CXLVI/3 (2000), pp. 437–51.
32 J. Hofmeister, M. Mihaljevič and J.Hošek, 'The Spread of Ash (*Fraxinus
    excelsior*) in Some European Oak Forests: An Effect of Nitrogen
    Deposition or Successional Change?', *Forest Ecology and Management*, CCIII
    (2014), pp. 35–47.
33 P. M. McEvoy and J. H. McAdam, 'Sheep Grazing in Young Oak *Quercus*
    spp. and Ash *Fraxinus excelsior* plantations: Vegetation Control, Seasonality
    and Tree Damage', *Agroforestry Systems*, LXXIV (2008), pp. 199–211.

34 J. Urban, J. Suchomel and J. Dvořák, 'Contribution to the Knowledge of Woods Preferences of European Beaver (Castor fiber L. 1758) in Bank Vegetation on Non-forest land in the Forest District Soutok (Czech Republic)', *Acta Universitatis Agriculturae et Silviculturae Mendelianae Brunensis*, LVI (2008), pp. 289–94.

3 The Mythology of Ash

1 Gary R. Varner, *The Mythic Forest: The Green Man and the Spirit of Nature* (New York, 2006), p. 39.

2 Åke Hultkrantz, *Prairie and Plains Indians* (Groningen, 1973), p. 8.

3 Kit Anderson, *Nature, Culture, and Big Old Trees: Live Oaks and Ceibas in the Landscapes of Louisiana and Guatemala* (Austin, TX, 2010), p. 99.

4 Kevin Crossley-Holland, *Norse Myths: Gods of the Vikings* (London, 2018), p. 24.

5 Darl J. Dumont, 'The Ash Tree in Indo-European Culture', *Mankind Quarterly*, XXXII/4 (Summer 1992), www.musaios.com, accessed 16 June 2019.

6 Jesse L. Byock, trans., *The Prose Edda* (London, 2005), p. 18.

7 Allaire K. Diamond and Marla R. Emery, 'Black Ash (*Fraxinus nigra* Marsh): Local Ecological Knowledge of Site Characteristics and Morphology Associated with Basket-grade Specimens in New England (USA)', *Economic Botany*, LXV/4 (2011), pp. 422–6.

8 Patricia Ann Lynch, *Native American Mythology* (Philadelphia, PA, 2010), p. 7.

9 Ibid.

10 Fred Hageneder, *Yew* (London, 2013), pp. 145–6.

11 Crossley-Holland, *Norse Myths*, p. 14.

12 Byock, *The Prose Edda*, p. 16.

13 Ibid., p. 120.

14 Crossley-Holland, *Norse Myths*, p. 187.

15 Byock, *The Prose Edda*, p. xix.

16 Ibid., p. 27.

17 Ibid., p. 18.

18 Anand Chetan, Diana Brueton and Allen Meredith, *The Sacred Yew* (London, 1994), p. 110.

19 Neville Fay, 'The Principles of Environmental Arboriculture', *Arboricultural Journal*, XXVI/3 (2002), pp. 213–38.

20 Harry Fokkens and Anthony Harding, eds, *The Oxford Handbook of the European Bronze Age* (Oxford, 2013), p. 22.

21 Crossley-Holland, *Norse Myths*, p. 123.

22 Douglas Forell Hulmes, *The Birth of 'Friluftsliv': A 150 Year International Dialogue Conference Jubilee Celebration* (Levanger, 2009).

23 Nora Chadwick and J. Corcoran, *The Celts* (Ringwood, 1970), pp. 28–33.

24 John Koch, *Celtic Culture: A Historical Encyclopedia* (Santa Barbara, CA, 2005), pp. 300, 421, 495, 512, 583 and 985.

25 W. Y. Evans-Wentz, *Fairy Faith in Celtic Countries* (West Valley City, UT, 2006), p. 390.

26   Varner, *The Mythic Forest*, p. 53.
27   Anna Franklin, *The Illustrated Encyclopaedia of Fairies* (London, 2004), p. 17.
28   Gabrielle Hatfield, *Encyclopedia of Folk Medicine* (Santa Barbara, CA, 2004), p. 16.
29   J. P. Frayne and Madeleine Marchaterre, *W. B. Yeats: Early Articles and Reviews Written between 1886 and 1900* (New York, 2004), p. 76
30   Jennifer Speake, *Oxford Dictionary of Proverbs*, 5th edn (Oxford, 2008), p. 233.
31   Donald Watts, *Dictionary of Plant Lore* (London, 2007), p. 15.
32   Ibid., p. 227.
33   Alexander Porteous, *The Forest in Folklore and Mythology* (New York, 2002), p. 93.
34   D. J. Conway, *Celtic Magic* (St Paul, MN, 2004), p. 134.
35   Varner, *The Mythic Forest*, p. 140.
36   Robert Graves, *The White Goddess* (London, 1948), p. 168.
37   Varner, *The Mythic Forest*, p. 53.
38   Danu Forest, *Celtic Tree Magic: Ogham Lore and Druid Mysteries* (Woodbury, 2014), p. 9.
39   J. Carney, 'The Invention of the Ogam Cipher', *Ériu*, XXII (1975), pp. 62–3.
40   Lloyd Robert Laing, *Orkney and Shetland: An Archaeological Guide* (Exeter, 1974), p. 134.
41   'A Literary and Philosophic Review', *Dublin University Magazine*, LXV (1865), p. 701.
42   Sandra Kynes, *Whispers from the Woods: The Lore and Magic of Trees* (Woodbury, 2006), p. 28.
43   Graves, *The White Goddess*, p. 264.
44   Stephen Fry, *Mythos: Greek Myths Retold* (London, 2017), p. 7.
45   Ibid., p. 21.
46   J. P. Lincott, *Myths and Legends of Flowers, Trees, Fruits and Plants* (Philadelphia, PA, 1925), p. 54.
47   Walter Keating Kelly, *Curiosities of Indo-European Tradition and Folk-lore* (London, 1863), p. 143.
48   H. Rackham, trans., *Pliny: Natural History*, vol. II (Cambridge, MA, 1940), p. 450.
49   V. Rydberg, *Teutonic Mythology*, trans. R. B. Anderson (London, 1889), p. 438.
50   Hugh G. Evelyn-White, trans., *Hesiod: Works and Days II* (Cambridge, MA, 1914), pp. 140–45.
51   Ibid., p. 109.
52   Marcel Detienne, *The Writing of Orpheus: Greek Myth in Cultural Context* (Baltimore, MA, and London, 1989), p. 42.
53   Evelyn-White, trans., *Hesiod*, pp. 140–45.
54   Jason Colavito, *Jason and the Argonauts through the Ages* (Jefferson, NC, 2014), p. 173.
55   Jennifer R. March, *Dictionary of Classical Mythology* (Oxford and Philadelphia, PA, 1998), p. 454.
56   Homer, *The Iliad*, trans. Stephen Mitchell (London, 2011), p. 460.
57   Forest, *Celtic Tree Magic*, p. 9.

4 Useful Ash

1 Oliver Rackham, *The Ash Tree* (Toller Fratrum, 2014), p. 21.
2 John Evelyn, *Sylva; or, A Discourse on Forest-trees, and the Propagation of Timber in His Majesty's Dominions* (London, 1664), p. 150.
3 William Cobbett, *The Woodlands* (London, 1825).
4 'Ash', www.botanical.com, accessed 16 June 2019.
5 P. T. Nicholson and I. Shaw, eds, *Ancient Egyptian Materials and Technology* (Cambridge, 2000), p. 341.
6 A. Bulleid and H. George-Gray, 'Glastonbury Lake Village', *Glastonbury Antiquarian Society* (1911), pp. 336–7.
7 Charlotte Erichsen-Brown, *Medicinal and Other Uses of North American Plants* (New York, 1979), p. 86.
8 J. P. Puype and H. Stevens, *Arms and Armour of Knights and Landsknechts in the Netherlands Army* (Utrecht, 2010), p. 343.
9 L. Allington-Jones, 'The Clacton Spear: The Last 100 Years', *Archaeological Journal*, CLXXII/2 (2015), pp. 273–96.
10 'Ash', Online Etymology Dictionary, www.etymonline.com, accessed 16 June 2019.
11 Ibid.
12 R. Marsden, *The Cambridge Old English Reader* (Cambridge, 2015), p. 460.
13 A. Verity, trans., *The Iliad* (Oxford, 2011), l. 165.
14 J. Evans, *The Ancient Bronze Implements, Weapons, and Ornaments of Great Britain and Northern Ireland* (New York, 1881), p. 313.
15 W. J. Sedgefield, *Beowulf* (Manchester, 1935), p. 151.
16 C. Williamson, *Beowulf and Other Old English Poems* (Philadelphia, PA, 2011), p. 151.
17 Francis Grose and S. Hooper, *A Treatise on Ancient Armour and Weapons* (Milton Keynes, 1970), p. 48.
18 Adriaan Beukers and Ed van Hinte, *Lightness: The Inevitable Renaissance of Minimum Energy Structures* (Rotterdam, 2005), p. 83.
19 Robert Morkot, *The A to Z of Ancient Egyptian Warfare* (Plymouth, 2010), p. 51.
20 Reginald Laubin and Gladys Laubin, *American Indian Archery* (Norman, OK, 1980), p. 21.
21 F. Hageneder, *Yew: A History* (Stroud, 2007), p. 104.
22 Ibid.
23 L.J.A. Villalon and D. J. Kagay, eds, *The Hundred Years War*, Part Two: *Different Vistas* (Leiden, 2008), p. 224.
24 R. Ascham, *Toxophilus: The Schole, or Partitions, of Shooting* (London, 1544).
25 Ibid.
26 E. Christiansen, *Norsemen in the Viking Age* (Oxford, 2008), p. 16.
27 Doris Gutsmied, Schümann Rheinische and Friedrich-Wilhelms, 'Merovingian Men: Fulltime Warriors? Weapon Graves of the Continental Merovingian Period of the Munich Gravel Plain and the

Social and Age Structure of the Contemporary Society – A Case Study', N-TAG TEN *Proceedings of the 10th Nordic TAG Conference at Stiklestad, Norway* (2012), p. 255.

28 John Seymore, *The Forgotten Arts: A Practical Guide to Traditional Skills* (London, 1984), p. 63.

29 Earle Clapp, 'Timber Depletion, Lumber Pieces, Lumber Exports and Concentration of Timber Ownership, Report on Senate Resolution 311', *U.S. Department of Agriculture and Forestry* (1920), p. 81.

30 John Robert Travis, *Coal in Roman Britain* (Oxford, 2008), p. 104.

31 Rackham, *The Ash Tree*, p. 54.

32 R. J. Mitchell et al., 'The Potential Ecological Impact of Ash Dieback in the UK', *Joint Nature Conservation Committee Report* (2014), p. 48.

33 Hilary Murray and Antoine Breandan Ó Ríórdáin, *Viking and Early Medieval Buildings in Dublin* (Dublin, 1983), p. 22.

34 Erichsen-Brown, *Medicinal and Other Uses of North American Plants*, p. 86.

35 Rackham, *The Ash Tree*, p. 57.

36 Ibid.

37 Bulleid and George-Gray, 'Glastonbury Lake Village'.

38 A. McMullen, R. G. Handsman and J. A. Lester, 'Nipmuc: A Key into the Language of Woodsplint Baskets', *American Indian Archaeological Institute* (1987), p. 170.

39 Sir G. Prance and M. Nesbitt, *The Cultural History of Plants* (New York, 2005), p. 318.

40 Margaret Hazen and Robert Hazen, *Keepers of the Flame* (Princeton, NJ, 1992), p. 159.

41 Alexa Dufraisse, 'Firewood Management and Woodland Exploitation during the Late Neolithic at Lac de Chalain (Jura, France)', *Vegetation History and Archaeobotany*, XVII/2 (2008), pp. 199–210.

42 'The Firewood Poem', www.allpoetry.com, accessed 12 April 2019.

43 'Wood as a Fuel: Technical Supplement', Forestry Commission England (2010), www.forestryengland.uk, accessed 9 April 2019.

44 J. B. Innes and J. J. Blackford, 'Palynology and the Study of Mesolithic– Neolithic Transition in the British Isles', *Archaeological and Forensic Applications of Microfossils* (2017).

45 Thomas Pennant, *A Tour in Scotland 1769 and a Voyage to the Hebrides*, vol. I (1772), p. 33.

46 L. Gustafsson and I. Ahlén, *Geography of Plants and Animals* [Statens lantmäteriverk, Svenska sällskapet för antropologi och geografi] (Stockholm, 1996), p. 158.

47 Elena Bargioni and Alessandra Zanzi Sulli, 'The Production of Fodder Trees in Morocco', *Ecological History of European Woodlands* (1996), p. 57.

48 European Wood Pasture (Silvopasture) Case Studies, www.coppiceagroforestry.com, accessed 18 July 2019.

49 'A Salute to the Wheel', www.smithsonianmag.com, accessed 8 November 2019.

50 P. T. Nicholson and Ian Shaw, *Ancient Egyptian Materials and Technology* (Cambridge, 2000), p. 341.

51 Ibid.

52 A. Shortland, 'The Social Context of Technological Change: Egypt and the Near East, 1650–1150 BC', Conference at St Edmunds Hall, Oxford, 12–14 September 2000, p. 63.

53 Thomas A. Kinney, *The Carriage Trade: Making Horse-drawn Vehicles in America* (Baltimore, MD, 2004), p. 34.

54 D. F. Austin, *Florida Ethnobotany* (Boca Raton, FL, 2004), p. 314.

55 W. J. Mills, *Exploring Polar Frontiers: A Historical Encyclopedia* (Santa Barbara, CA, 2003), p. 614.

56 Sean McGrail, *Boats of the World: From the Stone Age to Medieval Times* (Oxford, 2001), p. 191.

57 Robert Graves, *The White Goddess* (London, 1948), p. 263.

58 R. J. Hudson, 'Management of Agricultural, Forestry, Fisheries and Rural Enterprise – Volume II', EOLSS/UNESCO (2009), p. 116.

59 Stephen Wilkinson, *The Miraculous Mosquito*, www.historynet.com, accessed 19 July 2019.

60 Robert William Henderson, *Ball, Bat and Bishop: The Origin of Ball Games* (Urbana, IL, 2001).

61 'Louisville Slugger by the Numbers', www.slugger.com, accessed 19 July 2019.

62 Steve Craig, *Sports and Games of the Ancients* (Westport, CT, 2002), p. 84.

63 Green, *Wood: Craft, Culture, History*, p. 335.

64 Cameron Brown, *Wimbledon: Facts, Figures and Fun* (London, 2005), p. 16.

65 William J. Lavonis, *A Study of Native American Singing and Song* (Clunton, Shropshire, 2004), p. 31.

66 'Ash vs Alder: The Difference in Tone Woods Used in Fender', www.fender.com, accessed 18 July 2019.

5 The Healing Ash

1 I. Kostova and T. Iossifova, 'Chemical Components of *Fraxinus* Species', *Fitoterapia*, LXXVIII/2 (2007), pp. 85–106.

2 Iqra Sarfraz et al., '*Fraxinus*, A Plant with Versatile Pharmacological and Biological Activities', www.hindawi.com (2017).

3 'European Medicines Agency Assessment Report on *Fraxinus excelsior* L. or *Fraxinus angustifolia* Vahl, Folium' (2011), p. 10, www.ema.europa.eu, accessed 18 July 2019.

4 Gabrielle Hatfield, *Encyclopedia of Folk Medicine* (Santa Barbara, CA, 2004), p. 16.

5 John Gerard, *The Herball; or, Generall Historie of Plantes*, 1st edn (London, 1597).

6 Li Shizhen, *Compendium of Materia Medica: Bencao Gangmu*, ed. Luo Xiwen (Beijing, 2003).

7 You-Ping Zhu, *Chinese Materia Medica: Chemistry, Pharmacology and Applications* (Groningen, 1998), p. 208.

8 Daniel Moerman, *Native American Medicinal Plants* (Portland, OR, 1998), p. 528.

9 Ibid.

10 James W. Herrick, *Iroquois Medical Botany* (Syracuse, NY, 1995).

11 H. Rackham, trans., *Pliny: Natural History*, vol. IV (Cambridge, MA, 1968), p. 429.

12 Donald Watts, *Dictionary of Plant Lore* (London, 2007), p. 15.

13 Moerman, *Native American Medicinal Plants*, p. 528.

14 Charles F. Millspaugh, *American Medicinal Plants* (New York, 1974).

15 John Fiske, *Miscellaneous Writings of John Fiske*, vol. V (Boston, MA, 1902), p. 83.

16 'Assessment Report on *Fraxinus excelsior* L. or *Fraxinus angustifolia* Vahl, Folium', p. 10.

17 K. N. Venugopala, V. Rashmi and B. Odhav, 'Review of Natural Coumarin Lead Compounds for Their Pharmacological Activity', *BioMed Research International* (March 2013), Article id. 963248, p. 14.

18 I. Kostova, 'Review *Fraxinus ornus*', *Fitoterapia*, LXXII/5 (2001), pp. 471–80.

19 Jong Hoon Ahn et al., 'Secoiridoids from Stem Barks of *Fraxinus rhynchophylla* with Pancreatic Inhibitory Activity', *Natural Product Research*, XXVII/12 (2013), pp. 1610–14.

20 H. Wang, D. Zou and M. Xie, 'Antibacterial Mechanism of Fraxetin against *Staphylococcus aureus*', *Molecular Medicine*, X/5 (2014), pp. 2341–5.

21 Jong Hoon Ahn, Eunjin Shin, Qing Liu et al., 'Lignan Derivatives from *Fraxinus rhynchophylla* and Inhibitory Activity on Pancreatic Lipase', *Natural Product Sciences*, XVIII/2 (2012), pp. 116–20.

22 Darl J. Dumont, 'The Ash Tree in Indo-European Culture', *Mankind Quarterly*, XXXII/4 (Summer 1992), www.musaios.com, accessed 16 June 2019.

23 Rosa Guarcello et al., 'Insights into the Cultivable Ecology of "Manna" Ash Products Extracted from *Fraxinus angustifolia*, (*Oleaceae*) Trees in Sicily, Italy', *Frontiers in Microbiology*, X (2019), p. 984.

24 T. L. Luvisotto, R. N. Auer and G. R. Sutherland, 'The Effect of Mannitol on Experimental Cerebral Ischemia, Revisited', *Neurosurgery*, XXXVIII/1 (1996), pp. 131–8.

25 'Mannitol', www.scienceofparkinsons.com, accessed 27 July 2019.

26 Ibid.

27 'Report and Transactions', *Devonshire Association for the Advancement of Science Literature and Art*, VIII (1876), p. 54.

28 Roy Vickery, *Oxford Dictionary of Folk-lore* (Oxford, 1995), p. 18.

29 Margaret C. Ffennell, 'The Shrew Ash in Richmond Park', *Folklore*, IX/4 (1898), pp. 330–36.

30 Margaret Grieve, *A Modern Herbal*, vol. I [1931] (Mineola, NY, 2013), p. 67.

Epilogue

1 Oliver Rackham, *The Ash Tree* (Toller Fratrum, 2014), p. 15.

Further Reading

Austin, Daniel F., *Florida Ethnobotany* (Boca Raton, FL, 2004)

Byock, Jesse L., trans., *The Prose Edda* (London, 2005)

Cobbett, William, *The Woodlands* (London, 1825)

Crossley-Holland, Kevin, *Norse Myths: Gods of the Vikings* (London, 2018)

Culpeper, Nicholas, *Culpeper's Complete Herbal* (Manchester, 1826)

Edlin, H. L., *Woodland Crafts in Britain* (London, 1949)

Erichsen-Brown, Charlotte, *Medicinal and Other Uses of North American Plants: A Historical Survey with Special Reference to the Eastern Indian Tribes* (New York, 1979)

Evelyn, John, *Sylva; or, A Discourse on Forest-trees, and the Propagation of Timber in His Majesty's Dominions* (London, 1664)

Forest, Danu, *Celtic Tree Magic: Ogham Lore and Druid Mysteries* (Woodbury, 2014)

Frazer, James G., *The Golden Bough: A Study in Magic and Religion* (London, 1933)

Gaiman, Neil, *Norse Mythology* (London, 2017)

Graves, Robert, *The White Goddess* (London, 1948)

Grieve, Margaret, *A Modern Herbal*, vol. I [1931] (New York, 2013)

Hageneder, Fred, *The Living Wisdom of Trees* (London, 2005)

Loudon, John C., *Arboretum et fruticetum Britannicum* (London, 1838)

Mabey, Richard, *Flora Britannica* (London, 1997)

Miles, Archie, *Silva: The Tree in Britain* (London, 1999)

Mitchell, R. J., et al., *The Potential Ecological Impact of Ash Dieback in the UK*, JNCC Report 483 (Peterborough, 2014)

Penn, Robert, *The Man Who Made Things out of Trees* (London, 2015)

Petekin, G. F., 'Ash: An Ecological Portrait', *British Wildlife*, XXIV/4 (2013), pp. 235–42

Prance, Sir G., and M. Nesbitt, *The Cultural History of Plants* (New York, 2005)

Rackham, Oliver, *Ancient Woodland: Its History, Vegetation and Uses in England* (London, 1980)

—, *The Ash Tree* (Toller Fratrum, 2014)

—, *The Last Forest: The Story of Hatfield Forest* (London, 1989)

—, *Trees and Woodland in the British Landscape* (London, 1976 and 1990)

Stokes, J., and G. Jones, *Ash Dieback: An Action Plan Toolkit*, The Tree Council (London, 2019)

Strutt, J. G., *Sylva Britannica* (London, 1822)

Varner, Gary R., *The Mythic Forest: The Green Man and the Spirit of Nature* (New York, 2006)

Vickery, Roy, *Oxford Dictionary of Plant-lore* (Oxford, 1995)

Wallander, Eva, 'Systematics and Floral Evolution in *Fraxinus* (Oleaceae)', *Belgische Dendrologie Belge* (2013), pp. 38–58

—, 'Systemics of *Fraxinus* (Oleaceae) and the Evolution of Dioecy', *Plant Systematics and Evolution*, CCLXXIII (2008), pp. 25–49

Associations and Websites

ANCIENT TREE FORUM
www.ancienttreeforum.co.uk

ARNOLD ARBORETUM USA
www.aboretum.harvard.edu

THE ASH PROJECT
The Ash Project celebrates and protects ash trees and is an urgent response
to the effects of ash dieback on ash trees in the Kent Downs.
www.theashproject.org.uk

KEW ECONOMIC BOTANY COLLECTION
Online database collection of ash wood artefacts.
www.kew.org/science/collections/economic-botany-collection

MONUMENTAL TREES
An international inventory of big and old trees
www.monumentaltrees.com

PLANTS OF THE WORLD ONLINE
A multidimensional catalogue of plant life providing authoritative
information on the world's plant species
www.plantsoftheworldonline.org

TREE COUNCIL
Promoting a mission to protect the planet's trees and forests
https://treecouncil.org.uk

USDA FORESTRY SERVICE
www.fs.fed.us and www.fs.usda.gov

THE WOODLAND TRUST
The UK's Largest Woodland Conservation Charity
www.woodlandtrust.org.uk

Acknowledgements

I am very grateful to the many people who have helped and encouraged me. I must start by acknowledging the extensive work of Eva Wallander, R. J. Mitchell and Oliver Rackham, whose work has been invaluable to the research and writing of this book. I bow to their much greater expertise relating to *Fraxinus* and any mistakes in the presentation of the more complicated aspects of evolution, taxonomy, *Chalara* ash dieback and so on will be mine alone. Many other people have also provided general help and encouragement. The members of the Ancient Tree Forum – Ted Green, Jill Butler, John Smith, Brian Muleaner, Chris Knapman, Tim Hill, Russell Miller, Dave Clayden and many more – have assisted over the years with tree information and campaigning for ancient and veteran trees. The same is true of Nick Johannsen and Madeleine Hodge at the Ash Project. I would like to also thank Jon Stokes of The Tree Council who kindly helped me with information on the estimated number of ash trees in the British landscape and much more, and my friends and former colleagues at the Woodland Trust, particularly Nick Aitkin. Similarly, all my friends and colleagues at The Royal Botanic Garden Kew, particularly librarian Craig Brough, and Professor Donald Pigott for his help and information on the ancient pollarding systems of Cumbria and much more. Also to Bryan and Cherry Alexander – specialists in the study of peoples of the Arctic and the materials they use. I am grateful to all those experts who attended, and generously shared their research, at the Ashscape conference I organized at the Springhead Trust in 2016. These include Henry Kuppen, Director of Terra Nostra, for information on resistance to *Chalara* ash dieback in cultivars of *Fraxinus*, Vikki Bengtsson (Pro Natura – Sweden), Rob Wolton (Devon Ash Dieback Forum) and the artists Heather Ackroyd and Dan Harvey who created the giant ash tree art installation in the Kent Weald. I must also thank those people and organizations that generously helped with the illustrations. These include Eric Meier for the use of his ring pore image, Matt Heaton for use of his fossil image, The 'Weald & Downland Living Museum' for permission to photograph and use images (www.weald-down.co.uk), and The De Havilland Museum (www.dehavillandmuseum.co.uk). My thanks also to the board of trustees at Springhead, where I am Director, who have supported and encouraged me during the time I have taken to write

Ash: Nik Boulting, Kate Partridge, Ian Scott and Lee Smith. Thank you to my friend Fred Hageneder for suggesting to Reaktion Books that I should write this book. My thanks also to Michael Leaman and his team for all their patience and help.

Finally, I would like to extend my sincere thanks to my family: Edwina, Richard, Charles, Emma, Eppie and Aaron, and particularly Susannah Morris, who has supported me tirelessly throughout the long and painful process of researching and writing this book.

Photo Acknowledgements

᭥

The author and publishers wish to express their thanks to the below sources of illustrative material and/or permission to reproduce it.

FossilEra: p. 29 (Matt Heaton); Erik Meier: p. 117; Edward Parker: pp. 6, 8, 11, 12, 13, 16, 18, 20, 22, 24, 25, 32–3, 37, 38, 40, 42, 45, 46, 47, 52, 54–5, 57, 60, 62, 64, 66, 68, 69, 70, 73, 83, 85, 94, 98, 101, 102, 106, 114, 118, 119, 125, 128 (by kind permission of Weald and Downland Living Museum), 129 (by kind permission of Weald and Downland Living Museum), 130, 134, 135, 139, 142–3, 144 (by kind permission of Weald and Downland Living Museum), 147 (by kind permission of Weald and Downland Living Museum), 150, 151 (by kind permission of the de Havilland Aircraft Museum), 154, 158, 161, 167, 169, 170, 171, 172, 174, 178, 179, 180–81, 182, 183, 184; Shutterstock: pp. 10 (delcarmat), 35 (colin robert varndell), 36 (Wilfreda Wiseman), 41 (sarurun), 59 (Nigel Dowsett), 65 (DJTaylor), 71 (Karel Smilek), 72 (Herman Wong HM), 74 (K Steve Cope), 77 (Phil McDonald), 80 (Serge Goujon), 92 (Piotr Wawrzyniuk), 100 (Tntk), 103 (Thomas Males), 105 (ViMin), 122 (LGieger), 123 (Combatcamerauk), 132 (Ralph Eshelman), 146 (Nagib), 148 (Sissoupitch), 149 (trekandshoot), 155 (Eugene Onischenko), 157 (Roman Voloshyn), 162 (HelloRFZcool).

Index

❦